A MEANING TO LIFE

PHILOSOPHY IN ACTION
Small Books about Big Ideas

Walter Sinnott-Armstrong, series editor

Living with Darwin: Evolution, Design, and the Future of Faith
Philip Kitcher

Morality without God?
Walter Sinnott-Armstrong

The Best Things in Life: A Guide to What Really Matters
Thomas Hurka

Better than Human
Allen Buchanan

What's Wrong with Homosexuality?
John Corvino

The Character Gap: How Good Are We?
Christian B. Miller

On Romantic Love: Simple Truths about a Complex Emotion
Berit Brogaard

A Meaning to Life
Michael Ruse

A MEANING TO LIFE

Michael Ruse

OXFORD
UNIVERSITY PRESS

OXFORD
UNIVERSITY PRESS

Oxford University Press is a department of the University of Oxford. It furthers
the University's objective of excellence in research, scholarship, and education
by publishing worldwide. Oxford is a registered trade mark of Oxford University
Press in the UK and certain other countries.

Published in the United States of America by Oxford University Press
198 Madison Avenue, New York, NY 10016, United States of America.

© Oxford University Press 2019

CIP data is on file at the Library of Congress
ISBN 978-0-19-093322-7

9 8 7 6 5 4 3 2 1

Printed by Sheridan Books, Inc., United States of America

For Lizzie

Remember friend as you pass by
As you are now so once was I
As I am now so you shall be
Prepare for death and follow me

CONTENTS

Preface ix
Acknowledgments xiii

Introduction 1

1. The Unraveling of Belief 11

2. Has Religion Really Lost the Answer? 54

3. Darwinism as Religion 97

4. Darwinian Existentialism 133

BIBLIOGRAPHY 173
INDEX 185

PREFACE

This book is the end of a journey. For all that I have just quoted an uncomfortable epitaph that I saw once in a church in Norfolk in England, that sounds a little more sepulchral than I intend. It is true that I am approaching four score years—ten after my biblically allotted span. Although what I am saying is obviously connected to that, I am referring mainly to my academic career. The child is the father of the man. I was raised as a Christian—very intensely as a Quaker, with great stress not only on the moral duties of life but also on the mystical nature of the Godhead and of our encounter with It. Also, given that there are no creeds or dogmas, with the strong belief that one must work things out for oneself. As I did. Around the age of twenty, my faith faded, never to return. It would be easy to say that it was a direct function of starting into a life as a professional philosopher, reading Hume's *Dialogues*, and that sort of thing. Just not true, I am afraid. Natural theology was never a big part of my youth, and it is not now. I was a Kierkegaardian before I had even heard of Denmark. It was rather like baked beans and stamp collecting, two other great passions of my childhood. I just grew out of them. I did take up girls, which somewhat compensated.

My non-existent God was not about to give up that easily. I have a nasty feeling that He is a bit of a Calvinist who doesn't believe too strongly in the virtues of human free will. As I got under way as someone with a research agenda, I found that the issue of science and religion was always there to intrigue and excite me. In major part, this was because I became a historian and philosopher of science specializing in Charles Darwin's theory of evolution, past and present. As you can infer from this book, it was my love of Victorian culture that led me this way. I will never forget as a late teenager standing on the summit of Parliament Hill Fields, at the southern tip of Hampstead Heath, and looking down across London, marveling at that wonderful nineteenth-century structure, St. Pancras railway station. Darwin and his ideas were a natural, and it is hard to stay away from science and religion issues if this is the focus of your research. Factors from outside were also important. I got very much involved in the fight against the introduction of creationism in schools. This led me to the federal courthouse in Little Rock, Arkansas, where I was a witness for the American Civil Liberties Union, speaking on the natures of science and religion, Darwinian theory, and creationism (Ruse 1988a). Fighting alongside such people as paleontologist and popular-science writer Stephen Jay Gould, and America's leading Protestant theologian, Langdon Gilkey, was both a great responsibility and a great privilege. It was also, for someone with a personality like mine, great fun. I was writing and I continued to write on these topics, as I still do.

In recent years, I have felt more freedom to explore avenues that interest me—I am a full professor with tenure, I am not about to move, I have had all the fancy fellowships that one could get or could want to get, and, given that I am still working, my pension is in pretty good shape. I have gone in directions that upset many of my fellow Darwinians, most particularly looking at the similarities between our position and that of our opponents, Christians, often of the more evangelical kind. I am indeed rather proud of the fact that I have been called a "clueless gobshite" and that, in *The God Delusion* (2006), Richard Dawkins advised reporters always to check with someone else after they have talked to me. Jerry Coyne, the University of Chicago evolutionist, tells his readers that George Orwell was right; only an intellectual could believe the nonsense I propound.

Here, however, I do not want to talk of past triumphs. I will leave finding the sources for these comments as an exercise for the reader, but you might start with the somewhat overheated blog run by Coyne, "Why Evolution Is True." For some time now, I have sensed that I wanted to write not so much a synthesis but a capstone. Something that emerges in its own right from work that I have done. Something that asks whether the rather detailed and often inward-looking research to which I have devoted the scholarly part of my life—I never forget that my main role in life has been as a teacher—can speak to those ultimate existential questions about the meaning of life. Those questions that were so very important to me as a child. Perhaps that fading was more apparent than real. Truly, I still have a sneaking love of baked beans! Has my scholarly career been worth it? Has it taken my understanding forward? Do I have something to offer to others?

My answers are here in this book. My approach is often more personal than one expects in a book by a professional philosopher. I told you that I have reached the stage in my life where I feel more freedom! There is hardly a page where I have not drawn, often extensively, on previous writings. (Which is one excuse, I suppose, for so many self-references.) Although I am rather rude about some ways in which the notion of emergence is employed, this book is, in a very real sense, emergent. It comes from the past, but what emerges is here in the present, and it is new, not of the past. In a way, therefore, and this pleases me immensely, my book is truly evolutionary. In our physical structure, in our behavior, in just about every thought that we have, we humans show our past. We are something new and novel and unique. Are we better than the past? Is this book better than anything I have written before? That is for you to decide. You will find much in this book that is pertinent to these questions.

ACKNOWLEDGMENTS

The relationship between an author and his or her press editor is by its nature bound to be a little fraught. On the one hand, an author wants to get published. On the other hand, the editor wants to publish good books. These desires do not necessarily coincide. There is not exactly a master-slave relationship, but there is a teacher-pupil-like relationship. When it works, as in the classroom, it can be one of life's most wonderful experiences. Full of meaning! I have written or edited over sixty books now—I am still a long way behind Erle Stanley Gardner, the author of the Perry Mason mysteries—and I have had my share of editors. Overall, they have been wonderful people—as were my teachers—thoughtful, helpful, guiding firmly, and so pleased at my successes. Peter Ohlin, at Oxford University Press, is for me the Platonic Form of press editor. We met, as we often do, over lunch during one of my bi-annual trips to New York City to go to the Metropolitan Opera. I don't know quite whose idea it was that I should write this book, but I can say he has been with me all the way, right up to reading an earlier draft and pointing out gently that one can overdose on Michael Ruse's anecdotes about his remarkable past. He must be wrong there. There is no such thing as a surfeit of Michael Ruse stories. What, never? Well, hardly ever! The most important thing is that he saw I had something to say, and he let me say it in my own way. Thank you, Peter Ohlin. On the sale of the millionth copy, on my tab, we will split a jar of Quebec maple syrup.

Of the older people who have had huge influence on my life and thinking, there is none greater than my good friend and guide Edward O. Wilson. As a scientist, he is a major inspiration. Through his work both empirical and theoretical, no one more than he has brought greater honor to Darwin's theory. I regard *Sociobiology: The New Synthesis,* published in 1975, as visionary—both a telling of what has gone before and a program for what might come. More than that, Wilson has virtually from the first seen that we must take Darwinism from the laboratory and field study and into our lives, our understanding and ways of living. In this book, as you will gather, I disagree with many of the ways in which he wants to do this. Not all. He has been a major force in my rethinking about humans and their relationship with the rest of the living world. Most important, he has shown me that Darwinism matters not just to me as a historian but also as a philosopher. This book is a testimony to this.

In the same pattern of those who have inspired me, coming now from the humanities rather than the sciences, I very much suspect that my good friend Robert J. Richards, at the University of Chicago, will disagree strongly with both the argumentation and the conclusions of this book. He has done so for every book I have written thus far, so why should this be any exception? Nevertheless, like Ed Wilson, he, too, has long been a huge inspiration. Like me from a strong religious background—hard to imagine that Bob once thought of being a Jesuit, even to the extent of wearing one of those dark robes that make one look like an extra in a Verdi opera—he has, like me, fallen on secular ground. However, he has like me—or rather, I, like him—always thought that our shared love of Darwin and his revolution has to mean more than just history, however fascinating that may be. We have to see what it means to us in our lives and in our thinking about the point of it all: Meaning. (Interesting how when we get to these heavy-duty metaphysical questions, we feel the need to get into Germanic mode and capitalize.) Bob's answers are different from mine. Without presuming to put words into so distinguished a mouth, my hunch is that he is far more sympathetic to the approach I consider in chapter 3. I don't agree, but in disagreeing, I have learned much. You may even feel that I am in respects starting to edge closer to some of Bob's thinking. It is obviously the time for me to stop writing books before it is—time to slip into Germanic mode—TOO LATE.

Anyone who thinks that I take the science-religion relationship lightly has not read to the end of this book. Far more than to my fellow non-believers, I am in the debt of my Christian friends. Above all is Philip Hefner, long-time faculty member of the Lutheran School of Theology in Chicago and an ordained pastor, with whom, when we were younger, I used to spend a week every year at the conference of the Institute for Religion in an Age of Science, on Star Island off the coast of New Hampshire. He has a weakness for process theology, which I, as a very conservative non-believer, think heretical. Nobody's perfect. One doubts Phil believes in literal miracles and those sorts of things. I have never met anyone who walks more firmly in the footsteps of his savior, Jesus Christ. More recently, I owe much to Michael Peterson, my co-author of an Oxford book on science and religion—*Science, Evolution, and Religion: A Debate about*

Atheism and Theism. A Methodist theologian and philosopher, if he is as good in the classroom as he was in teaching me, his college (Asbury Seminary in Kentucky) has a jewel indeed. John Kelsay, my colleague here at Florida State University, a student of Islam and a Presbyterian minister, has been a friend and guide and comfort for the last two decades. It is a great privilege to team-teach with him. Last but not least—never least—are my former graduate students Dan Deen and Elizabeth Jones. I had a rotten time making the transition from the rather close environment of a Quaker boarding school to university life, and my mission has always been (and still is) with first- and second-year undergraduates. It is only more recently that I have become truly engaged at the graduate level and discovered the rich rewards that such contact brings. Of Dan and Liz I will say simply that, deeply committed Christians, more down-to-earth human beings I have never met. Makes me think.

Nearer to home, Jeff O'Connell, my beloved, now graduated doctoral student and research assistant, has helped me (who am totally technologically inadequate) find the references and books that I needed. Contrary to the suspicion of some, I am not like the leadership of the United States of America, where every thought comes totally uninfluenced by outside sources. When you read this book, you can guess which of the copies of Roy Lichtenstein's works I have left Jeff in my will. As always, Lizzie—to whom this book is dedicated—is the light of my life, as she has been since I first met her nearly forty years ago. I loved her then, and I love her now. This despite her lack of appreciation for my taste in breakfast spreads. I am sure that Peter Ohlin will let a fellow Canadian share the maple syrup. And then there is Scruffy McGruff. Although he enters late, he really is the star of this book. I will let him speak for himself. I might as well. Just like the author, he does so non-stop. He still cannot quite understand why the book is not dedicated to him.

A MEANING TO LIFE

| INTRODUCTION

You are born. You live. Then you die. It's terrifying. If you don't think so, then you should! We came from an eternity of oblivion. We return to an eternity of oblivion. In between? We try to make sense of it all and to lead worthwhile lives. In the end, you know truly that it doesn't mean a thing. Shakespeare (in *Macbeth*) is good on this:

> And all our yesterdays have lighted fools
> The way to dusty death. Out, out, brief candle!
> Life's but a walking shadow, a poor player
> That struts and frets his hour upon the stage
> And then is heard no more: it is a tale
> Told by an idiot, full of sound and fury,
> Signifying nothing.

The great eighteenth-century Scottish philosopher David Hume was the arch-skeptic. His delving into the nature of knowledge led him to the conclusion that there is no absolute support for anything. We think the sun will rise tomorrow, but, as

Bertrand Russell was to say sardonically, we could be in the state of the turkey on Christmas Eve. We expect that nice farmer to come along with a bucket of feed. Oh, what's this? Why is he carrying an ax instead? Fortunately, said Hume, our psychology protects us from the hole into which philosophy has pitched us. "I dine, I play a game of backgammon, I converse, and am merry with my friends. And when, after three or four hours' amusement, I would return to these speculations, they appear so cold, and strained, and ridiculous, that I cannot find in my heart to enter into them any farther" (Hume 1739–40, 175).

Are we not in the same place with respect to our worries about meaning and what we might call the "ultimate questions"? If we think about it, if we use our reason, then we have no answers. To say it again, it is terrifying. Tolstoy had it right. Everything we do and say ends as "stench and worms" (1882, 7). Obviously, though, we couldn't go through life, we couldn't function as organisms, if this reality was ever before us, paralyzing our will, and unable to do anything. Our psychology must and does rescue us. My wife and I take the kids to the seaside. It's fun. The children have a good time. My wife and I have the added satisfaction of working as a team, making for a functioning and healthy family. At one level, obviously, it is all part of biology. More little Ruses coming into the gene pool to ensure the future. This does not deny that at another level, it all makes good sense, with lots of meaning and purpose. A day out like this is as far from "signifying nothing" as it is possible to be. We know what signifying nothing is all about. It is a day spent at home, vegging out, non-stop watching television of the lowest denominator. The soaps and news programs about Dear Leaders.

Fair enough. But let's not kid ourselves; in the end, the problems of us little people "don't amount to a hill of beans in this crazy world." The American Pragmatist philosopher William James knew the score, admitting that "we are all such helpless failures in the last resort" (1902, 47). He laments the fact that the "sanest and best of us are of one clay with lunatics and prison inmates, and death finally runs the robustest of us down." No wonder then that after reflections like these, "such a sense of the vanity and provisionality of our voluntary career comes over us that all our morality appears but as a plaster hiding a sore it can never cure, and all our well-doing as the hollowest substitute for that well-BEING that our lives ought to be grounded in, but, alas! are not." The only consolation we have is, as the little epitaph told us, we are all in it together. No one is going to escape. Not the president. Not the queen. Not the dean of arts and sciences. We are all going to die, and with death come darkness, stillness, nothing.

Close this book before we start. Well, not quite so fast! Many wise and sensible, decent people—in the past and still in the present—think we are ignoring the crucial element in the story. Life is not meaningless, as we have been suggesting. It all makes very good sense, in general and in particular for us. You simply have to accept that there is a Creator, an intelligent force, a God behind all of this. William James, having dug the hole, now fills it—"here religion comes to our rescue and takes our fate into her hands" (1902, 47). This fills us with joy: "what we most dreaded has become the habitation of our safety, and the hour of our moral death has turned into our spiritual birthday. The time for tension in our soul is over, and that of happy relaxation, of calm deep breathing, of an eternal present, with no discordant future to be anxious about, has arrived. Fear is not held in abeyance as it is by mere morality, it is positively expunged and washed away."

For people in the Abrahamic traditions, this supposed deity is not just any old god. It is God—Yahweh, Jehovah—the God of the Old Testament. "Which made heaven, and earth, the sea, and all that therein is: which keepeth truth for ever" (Psalm 146:6). For the Christian, there is Jesus the Messiah, who died on the Cross for our eternal salvation. Those writing on the meaning of life stress again and again that this God is the way and the only way. In the words of the Protestant writer William Lane Craig, if there is no God, the world is "the result of a cosmic accident, a chance explosion. There is no reason for which it exists. As for man, he is a freak of nature—a blind product of matter plus time plus chance. Man is just a lump of slime that evolved into rationality" (2000, 45).

What exactly is this God that gives our lives such meaning, and what is our place in the equation? As we philosophers like to say, obviously a lot is going to depend on what you mean by "meaning." We'll be working on this one through the whole book. For now, let us agree that it is something of value, of worth, to be desired. I'm not sure that there is an identity. Is everything of value that meaningful? A game of golf? Is everything meaningful that valuable? Adolf Hitler? But we can surely say that the Christian horror at the alternative, nihilistic alternative—there is nothing— is bound up with value. The alternative means that nothing we do has any value or worth. Shakespeare was right about us just being walking shadows. What then is the value in or from God? His very being, for a start. He is perfect, without cause, totally good, all-knowing, and all-powerful. If this is not what we mean by valuable and meaningful, then what is? We humans get our meaning and value from the fact that we are created by this God, out of love, for Him to cherish and for us to adore and obey Him. Christians, of course, have their own particular take on all of this. We were created perfect, but then we fell, and God Himself had to sacrifice on the Cross, but in so doing, He put things back together again, and we can again continue eternally in our symbiotic relationship with our Creator.

If you put things this way, it all seems pretty cut and dried. Given the comforts of Christianity, who but a total fool or lunatic would opt for anything else? Why would you be like the unfortunate clergyman Mr. Prendergast in Evelyn Waugh's *Decline and Fall* and have "doubts"? If you do have them, then why not follow the advice of Pascal and push them down? After all, if there is something, you win, and if there is nothing, you hardly lose. In any case, if you try hard enough at believing, you may end up believing after all. And yet people do. Have doubts, that is. Actually, not so much doubts as absolute convictions that what you see is what you get, and it isn't necessarily all that nice or meaningful in any sense of the word.

The nineteenth-century German philosopher Arthur Schopenhauer was good on this sort of thing. Here is a characteristically paragraph-long sentence on the topic. "If you try to imagine, as nearly as you can, what an amount of misery, pain and suffering of every kind the sun shines upon in its course, you will admit that it would be much better if, on the earth as little as on the moon, the sun were able to call forth the phenomena of life; and if, here as there, the surface were still in a crystalline state" (1851, 46). Warming to his theme and turning to us humans: "If children were brought into the world by an act of pure reason alone, would the human race continue to exist? Would not a man rather have so much sympathy with the coming generation as to spare it the burden of existence? or at any rate not take it upon himself to impose that burden upon it in cold blood."

Schopenhauer had his twentieth-century counterparts. There was the French existentialist Albert Camus, for instance, assuring us that there "is but one truly serious philosophical question, and that is suicide" (1942, 3). Existence is absurd. We are in the position of Sisyphus, forever rolling the rock up the hill, only to have it roll back down and having to start again. "It is legitimate to wonder, clearly and without false pathos, whether a conclusion of this importance requires forsaking as rapidly as possible an incomprehensible condition" (6). Moving on to the twenty-first century, the South African philosopher David Benatar can give the Germans and the French a run for their money. "We are born, we live, we suffer along the way, and then we die—obliterated for the rest of eternity." What's it all about? Benatar concludes sagely and fearlessly, "ultimately nothing." Continuing to the dreadful conclusion: "Despite some limited consolations, the human condition is in fact a tragic predicament from which none of us can escape, for the predicament consists not merely in life but also in death" (2017, 12).

There you have the answer, and there for me you have the question. If life is so awful, why then do we have the Schopenhauers and Benatars of life, or, perhaps more appropriately, of existence? Why don't we all just take the God alternative, the Christian alternative, perhaps, and have done with it? Why indeed keep on harping about it all? Why not drop the topic and move on? Simply because that is what philosophers do, and I am a professional philosopher, and that is what I do. Say it again, Socrates! "The unexamined life is not worth living." Caveat emptor. I am truly a historian of ideas and not really that immersed in the usual "analytic philosophy" of today's anglophone thinkers. Until starting the project leading to this book, I had not realized quite what a cottage industry meaning-of-life discussions had become. I am (to say it again) a professional, and you will see that I take note of and use (with appreciation and gratitude) ideas from this industry. I have already started to do so. However, said he rather proudly, you will see that I do not fit the usual pattern.

My quest on the meaning-of-life question began independently about twenty years ago, when I was invited to give some of the Gifford Lectures (Ruse 2001b). Explicitly, these are intended to "promote and diffuse the study of natural theology in the widest sense of the term—in other words, the knowledge of God." Hence the spark that led me here, for I am an ardent Darwinian evolutionist, and it was this that informed my lectures. I think that the implications of Charles Darwin's ideas, as given in his *On the Origin of Species* (1859) and *The Descent of Man* (1871), are the most important things in Western thought, at least back to the Copernican Revolution. Specifically, I think the fact that we humans are the end products of the slow, law-bound process of natural selection has huge implications—huge implications for questions about the meaning of life. You want to talk about the knowledge of God? Talk about Darwin. That was my contribution to the theme of 2001's Gifford Lectures, and that continues to be my stance today.

If further justification is needed in taking up a much-discussed topic, it is that this all-important event barely, if at all, makes it into the analytic discussions of the meaning of life. To be fair, although there are many who deny Darwin's contribution, people are not creationists. They just don't think that evolution, including Darwinian evolution, is all that important. Had not the great Ludwig Wittgenstein in his groundbreaking *Tractatus* said, "Darwin's theory has no more to do with philosophy than any other hypothesis in natural science"? He kept up this theme. Around 1950, not long before his death, he was still saying, "I have often thought that Darwin was wrong: his theory doesn't account for all this variety of species. It hasn't the necessary multiplicity." (Ruse (2017b) discusses these and related philosophy-originating, Darwin-flogging views and gives references.)

I am not going to argue the toss here or try to interpret Wittgenstein's somewhat gnomic remark. Just make my own case positively. The key will be the actual scientific claim about evolution through natural selection and its application to us humans. For now, though, I want to pick up on Darwin's methodology, namely seeking information in the past to throw light on the present—for instance, about why the birds of the Galapagos archipelago are like those of South America and why the birds of the Canary Islands are like those of Africa. I want to start by turning to history and what seems to me to be the biggest puzzle. A puzzle of a personal nature. I have recently edited two collections on atheism, its nature and its history (Bullivant and Ruse 2013; forthcoming). One thing that has struck me and my very nice Christian co-editor is how hard it is to find anyone to write on the Middle Ages. We just cannot seem to ferret out atheists anywhere. Everyone was in the Christian paradigm, as one might say—or a Jew or a Muslim—and there was value and meaning everywhere. Yet after that, as time goes by, our job gets easier and easier—there are God-deniers everywhere—until the biggest problem is saying what we don't want included!

That is where I want to start, then, with the history of the decline of God-belief and the increasing willingness to embrace a nihilistic view of existence. As you might expect, toward the end of this history, Darwin's theory is going to appear. A big part of the discussion will be on the theory's nature and implications—or not, as the case may be. Against this historical background, I shall turn to philosophizing, and I shall structure the discussion on a very helpful taxonomy embraced by the analytic philosophers (Metz 2013). First, I shall turn to theism and see where things stand today in the light of history, in the light of Darwin. I shall avail myself of what I think is a very useful dichotomy: meaning in light of the God question, meaning in light of the human soul question. To this end, I shall supplement the discussion of Christianity—for this will be my main religion—with a brief discussion of Buddhism, which, as is well known, has a rather different take on the God question but when it comes to us has strong feelings about meaning.

I move second to non-theism—let us call it naturalism—and again I shall invoke an established dichotomy, this time between objective meaning or value and subjective meaning or value. I shall show that Darwin has been—still is—taken in both directions. I shall also show why, for me, the subjective route is the one to take. Yet, although I am so far with the naysayers as to think of myself as a Darwinian existentialist, I shall argue that there is more to the story than Schopenhauer and his fellows allow. I am not in the suicide business! I am not sure that Camus was, either, however much he seemed to enjoy wallowing in the absurdity of it all. Whatever. I am led back to Hume and what he had to say on skepticism and psychology. Which will bring me happily to a very Darwin-inspired concluding meditation on how limited is our knowledge about any of it. A message that all philosophers confront at some point or another.

THE UNRAVELING OF BELIEF

Medieval Religion

I want to start in England around 1500, the end of the fifteenth century and the beginning of the sixteenth (Duffy 1992). I choose England because that was the country of my birth and my up-bringing and the one whose history I know best. I am sure someone living in one of the great countries in Western Europe—France and Germany, particularly—could write something similar. If you look just at the bare statistics, it all seems grim. Today, life expectancy in England is around eighty years—a bit more for women than for men, a bit more in the more prosperous south of the country than in the north. If you go back five hundred years or so, however, life expectancy is less than half of that—around thirty-five years, to be exact. That, however, is a little misleading, for up to a third of children didn't make it past the first year, and a tenth or more of women died in childbirth. If you got past these hurdles, you might make it to around forty-five. Of course, as today, some people lived much longer, but they were the exceptions.

There is no big mystery here. As the seventeenth-century English philosopher Thomas Hobbes said, life tended to be "solitary, poor, nasty, brutish, and short" (1651, Chap. XIII, "Of the naturall condition of mankind"). He said it in the context of the Civil War, but it applied in peacetime, too. Certainly, if you were a member of a family that was prosperous, say, a rich merchant in a town like Norwich in East Anglia—making a fortune out of the wool trade—there were going to be compensations. These would be limited and restricted. You got breast cancer, and you were doomed. The same if you fell off your horse. That is before you start on the various diseases and plagues. Not to put too fine a point on it all, everyone lived every day with a background stench of fecal matter—if not that of you and your family and your servants, then that of your animals. Today, we use water to flush it all away. Then, as like as not, you drank the water. These were the rich and comfortable. Think of the laborer and his family in their cottage. Cold, hungry, ignorant, tired all the time, dirty—the list is endless. Seemingly, Schopenhauer had a point.

There was another side to it all, making if not for an outrightly happy life, then certainly a life of meaning, one that gave positive joy and freedom to nigh everyone, prince and pauper (to use a cliché). This was religion, more particularly, the Catholic version of Christianity. This had both an inward side and an external side, although truly they were one. The inward side was the traditional story of Christianity. The world and everything in it was created by a loving God. Humans had a special place—we were "made in the image of God"—but we fell, and hence something was needed for our eternal salvation. This could only be done by God Himself, so He came to earth in the form of Jesus Christ, and through the sacrifice on the Cross, all was again made possible. We can join God in heaven, but only if we behave down here on earth.

To this traditional story was added a great deal of comforting detail—virtually all, as the future Protestants were to point out, without biblical support (MacCulloch 2004). Most obviously, there was the status and role of the mother of Christ, the Virgin Mary. To put it candidly, the biblical evidence for this is thin to non-existent. Not all the Gospels dwell on her early experiences of giving birth to the Messiah, and sometimes the move is the other way. "Who is my mother? and who are my brethren?" (Matthew 12:48). The answer comes: "For whosoever shall do the will of my Father which is in heaven, the same is my brother, and sister, and mother" (12:50). Well, yes, but . . . Yet, to quote a contemporary discussion, "Mary, also known as St. Mary the Virgin, the Blessed Virgin Mary, Saint Mary, Mary Mother of God or the Virgin Mary is believed by many to be the greatest of all Christian saints" (Catholic Online). The Virgin Mother "was, after her Son, exalted by divine grace above all angels and men." She was not alone. She was not just the greatest, she was the greatest of a great many Christian saints. Most prominently, of course, were the great biblical saints, like Saint Peter, Saint Paul, and Saint Stephen, the first Christian martyr. These belonged to everyone. Then there were the national saints. In England, above all others, was Saint Thomas à Becket, who was murdered in 1170 by followers of King Henry II on the altar steps in Canterbury Cathedral. Also, there were regional saints as well as those to whom one prayed for help on specific matters. Saint Apollonia had all her teeth pulled out or smashed when she was tortured for her faith. Naturally, it was to her that one turned with a toothache.

Already we start to get to the external side of Christianity around 1500. The various doings of Jesus and the saints were very much a two-way business. You could pray to Saint Apollonia for help, but she and the others expected something—respect, celebration, proper conduct—in return. Above all, there was Easter, when our Lord made the ultimate sacrifice. You are just coming to the end of forty days of avoiding meat—serious fasting in some cases—and now the tragedy and the triumph. Celebrating Jesus's ride into Jerusalem on a donkey, Palm Sunday was a day of many processions. In England, to mark the throwing of palms to soften his way, there was a major blessing of substitutes, like the leaves of the yew. (In my childhood, instead of going with my parents to Quaker meeting—which religion rather primly refused to acknowledge any church festivals—on Palm Sunday, my grandmother was allowed to take me to a high Anglican church, where we were given palm leaves twisted and tied into a cross.) Then there were the rituals of Maundy Thursday—the queen still hands out money to the aged poor—Good Friday and ceremonial buryings of the host, and finally Easter Sunday and joyful hymns of thankfulness and praise.

This was the highlight of the Church year, but there was always something to keep one busy. In Long Melford, the rich wool town in Suffolk, bonfires were lit to celebrate Midsummer Eve (June 24, Saint John's Day), the eve of Saints Peter and Paul a week later, Saint Thomas of Canterbury (Thomas à Becket) another week later, and finally, on July 25, Saint James's Day. The rich took the opportunity of scoring one with God by providing feasts for the poor—ale, mutton, and pease puddings. (Only in England! Split peas and bacon.) Lives were full of meaning, both here and now and in the future. This was explained and reinforced by sermons, backed by the illustrations and objects of devotion in church, the processions and the costumes, and often-enacted dramas to illustrate significant events in biblical history. The York Mystery Plays, done by various guilds on carts that moved around the town, were the highlight of the celebrations of Corpus Christi, around the beginning of June. Suppressed during the Reformation, they were resurrected after the Second World War. My one claim to thespian fame is that in 1957, in that year's cycle, I had a very minor role as a scene-shifter. Increasingly, it seems that was the only time I shall ever appear on the stage with Dame Judi Dench, a then-unknown young professional who was playing the Virgin Mary.

Above all, there was meaning at the moment of transition from one state to another. Everyone remarks on the religious signifi-cance of death in medieval Europe. The end of life was the social event of one's whole existence. Many were the prayers to make sure that one did not die alone but rather with priest and family and friends. An even greater invention than the elevation and exalta-tion of the Virgin was the idea that not only do the damned go down to hellfire and torment, but the saved also have to do a span of rather unpleasant penance to pay for their sins. Purgatory was big time. As we all know, it was the focus of the criticisms of the Reformers, Martin Luther particularly, for it was possible to buy off the extent of one's time in Purgatory. Masses and gifts to the Church were thought particularly efficacious, as were other things like supporting alms houses for the respectable, aged indigent. (The plot of Anthony Trollope's first Barchester novel, *The Warden*, written in the 1850s, revolves around one such establishment.)

It is important to emphasize that dying and making preparations for the future—your own soul and those of others still living—were not all gloom and doom. The very point was that there was continuity. Especially with respect to meaning. Purgatory was not much fun. It all made good sense, both for the future and for the present. Just as you were expected to provide for your immediate family, so you were expected to provide for the needy and unfortunate. That was what being a Christian was all about. When Christianity was functioning properly, life made a lot of sense. For here and now and for the future.

The Story Comes Apart

What went wrong, if that is the right term? It is no great secret. The three Rs: the Renaissance, the Reformation, and the (Scientific) Revolution. Before we attempt an integrated picture, let's take them in turn. The Renaissance, starting at least a century before the full activity of the sixteenth century, saw an invigorating of many aspects of human life and culture—in the arts, in music, in politics, and more. For us, all-important is the discovering afresh the glories of Greece and Rome, particularly in their extant writings, something made more accessible by a parallel interest in the languages of Greek and Latin and the desire to know them and use them effectively. The paradigm, of course, was Erasmus of Rotterdam, most particularly his preparing of high-quality editions and translations of the Bible.

Turning to Greece and Rome meant turning to non-Christian pagan times, including those of the great philosophers Plato and Aristotle. Neither Plato nor Aristotle was remotely atheistic—Plato had the Good, and Aristotle had his Unmoved Mover—and their ideas had long been incorporated into Christian thinking. Although Augustine did not read Greek, through the writing of the Neo-Platonist Plotinus, he knew the full details of the Platonic philosophy and embedded much of it into his innovative theology. Nearly a thousand years later, Thomas Aquinas likewise was no Greek reader, but by the time he was writing, Aristotle had been recovered and translated, and that transformed Aquinas's thinking profoundly. However, say what you like, both Plato and Aristotle were offering world pictures with no mention of Jesus Christ (let alone Adam and Eve, who made it all necessary), and their idea of God was (at most) but part of the Christian idea of God. More on this point later, for here I want to stress that whatever the issues with these philosophers, it all pales beside some of the stuff being unearthed by the humanists of the Renaissance. Above all, there was the Epicurus-influenced poem *De rerum natura* (*On the Nature of Things*) by the first-century B.C. Roman poet Lucretius. In the tradition of the at-omist Democritus, this pulled no punches. It's all a happy accident.

> At that time the earth tried to create many monsters
> with weird appearance and anatomy—
> androgynous, of neither one sex nor the other but some-
>   where in between; some footless, or handless;
> many even without mouths, or without eyes and blind;
> some with their limbs stuck together all along their body,

and thus disabled from doing harm or obtaining anything
    they needed.
These and other monsters the earth created.
But to no avail, since nature prohibited their development.
They were unable to reach the goal of their maturity,
to find sustenance or to copulate.

At first, everything is a non-functioning mess. Infinite time cures
all, and things started to work.

First, the fierce and savage lion species
has been protected by its courage, foxes by cunning, deer by
    speed of flight. But as for the light-sleeping minds of dogs,
    with their faithful heart,
and every kind born of the seed of beasts of burden,
and along with them the wool-bearing flocks and the horned
    tribes,
they have all been entrusted to the care of the human race.
    (5.862–867)

Obviously, the discovery of things like this did not mean that the
good citizens of Long Melford gave up Jesus and the Virgin, lost all
sense of life's meaning, and stopped providing pease pudding for
worthy plowmen and their dames. It was nevertheless out there in
the open, dripping away like water on a rock.

The Reformation meant many things. What it certainly did
not mean was an end to Christianity and the sense of purpose.
Anything but! In many, many respects, people like Martin Luther
and John Calvin were more sincere than the Roman prelates and

priests they were displacing. There was a whole new emphasis on the truth of the Bible, rather than the authority of the Church. Although not every reading was literal—Luther, for instance, was not at all keen on the Epistle of James, because he thought there was too much emphasis on good deeds—there was a drive to a simpler, more straightforward understanding, unencumbered by all the metaphorical interpretation that scripture had acquired. Above all, there was a hatred—that is not too strong a term—of the idea that good works at all influence or determine our future fate. In a return to Augustine, there was a total emphasis on faith and grace. Accept Jesus, and you have a chance of being in. Don't accept Jesus, and you are certainly out. Even more certainly out was Purgatory. Apart from its lack of biblical support, it was seen by the Reformers as a trick to con the faithful into generous support of the Church. With the emphasis on faith and faith alone—*sola fidelis*—paying for masses and the like was simply a waste of time.

Showing how artificial it is to separate the Renaissance from the Reformation, the emphasis on the Bible and de-emphasis of the Church meant that laypeople needed direct access to the Holy Word. Thus, people like Luther set about offering editions of the Bible in the vernacular, and to do this, the groundbreaking work of people like Erasmus—who never left the Mother Church—was essential. Printing helped, too, of course. No longer were copies of the Bible handwritten with the consequent opportunities for mistakes. As with the Renaissance, though, there was a down-side, if not at first, then later. In the West, no longer did you have just one version of Christianity—there was always the Eastern version—but now you had Catholic and Protestant and, before long, Protestant and Protestant and Protestant. Did the bread and wine still have Jesus present, as is claimed by Lutheran consubstantiation, or is it just symbolic, as it was for Calvin? Can you take up arms, as is insisted by Lutheran and Calvinist, or must you be a pacifist, as is insisted by the Anabaptists? With so many options on offer, thoughts of jettisoning it all started to lurk. At least, thoughts of a world without Jesus.

Don't exaggerate. For Protestants, Jesus and Jesus alone—kick out the Virgin and most of the saints—was the key to salvation. Augustine was the guide here. Worried about the need for Jesus to die on the Cross, he found the solution in the Old Testament. Adam and Eve fell, and because of this, we are all tainted with sin. Salvation comes through the sacrifice of Jesus (God) for our sakes. He is the innocent lamb who atones for our sins. (There were debates about whether He removed the sins or simply declared us sinless.) Although as historian Peter Harrison (2009) has pointed out, this somewhat paradoxically leads us from the Reformation to the Revolution and another source of anxiety and loss of faith. If you are stressing original sin, then you are stressing the Fall, and that leads naturally to questions about Adam's prelapsarian state. Made in the image of a good God, he had to be super-intelligent! After all, he would not have been able to name the animals had he not been. Why then are we not equally super-intelligent? Because of original sin! It messes with us epistemologically as much as it messes with us ethically. (As we shall see in chapter 2, this is still the position of the Catholic Church.) We cannot at once know the truth, but as part of our God-given duty, we must strive to be good, so we must strive for empirical understanding. We will never get there. It will not be enough, but that does not deny our duty. It is easy to see how this proto-Popperian philosophy of striving pushed people toward science and its understanding. Not the only factor kickstarting the Scientific Revolution but a significant one.

In major respects, the Revolution in science was akin to what was going on in the Renaissance and in the Reformation—major changes in which one can see glimmerings of non-belief but far from anything definitive or intended (Hall 1954; Hall 1983; Dijksterhuis 1961). Copernicus, a Polish Catholic, was a minor cleric who died in good standing with his Church. But when he moved the sun to the center of the universe and demoted the earth to planet status—one of several—going around the sun, he was going against Aristotle, with reason now considered one of the major supports of Catholic theology (Kuhn 1957). Obviously, also, we just no longer seem quite so important. What about all those stars? Wasn't it reasonable to think of them as suns, and if so, wasn't it reasonable to think of them as also having planets? And if there were planets, then surely God wasn't going to waste them? Little wonder that we find people speculating on the "plurality of worlds" (today known as "cosmic pluralism") question or problem. It was not always safe to do this, and at the end of the sixteenth century, the Dominican friar Giordano Bruno got burned at the stake for such speculations.

None of this cosmological speculation was disproof of the existence of God. From a natural theological perspective—reason and evidence rather than faith alone (revealed theology)—one could with reason say one had ever-increasing evidence of God's design. Even life elsewhere could be taken as proof of God's designing powers, although there were dangers here. Perhaps no other intelligent beings are fallen as we are, but if some are, does this mean that Jesus has to be sacrificed on a weekly basis all over the universe?

Kind of takes away from the significance of the drama down here if, even as I write, on some planet within Andromeda, He is being nailed to the Cross (Ruse 2001a).

Root Metaphors

There was one major striking thing that happened during the Scientific Revolution, and turning to it guides us toward a unifying explanation of what was happening in the sixteenth and seventeenth centuries (Ruse 2017b). Aristotle had divided causation into a number of categories, the most important of which were what we would call "efficient causes" and what even Darwin—especially Darwin—referred to as "final causes." On the one hand, we have the physical (or mental) things that make other things happen. I hammer on a nail, and there is sound, and the nail is firmly embedded in the wood. On the other hand, we have purposes or intentions. I hammer the nail in order to build a house, an end that I value. Note that these two kinds of causes, for all that one refers only to the past and the other brings in the future, are not entirely symmetrical. If the nail is in the wood, then someone did some hammering. If the nail is in the wood, it does not follow that the house will be completed. I might get a job on the other side of the country and the house never gets finished. Its frame is torn down for its lumber. A big question now was whether final causes were legitimate or should be done away with. Are they a sign of intellectual weakness? Increasingly, the feeling was that this was true in the physical sciences at least. Francis Bacon likened final causes to Vestal Virgins, decorative but sterile. The great French thinker René Descartes (1596–1650) argued that ontologically, God created two basic substances: *res extensa* and *res cogitans*, things extended and things thinking. The mark of the material world is that it has spatial dimensions. It is completely inert, unthinking, basic. Ends and values are precisely the sorts of things that *res extensa* cannot have. In any case, Descartes noted (perhaps somewhat disingenuously), one could never be quite sure what end God intended: "there is an infinitude of matter in His power, the causes of which transcend my knowledge; and this reason suffices to convince me that the species of cause termed final, finds no useful employment in physical [or natural] things; for it does not appear to me that I can without temerity seek to investigate the [inscrutable] ends of God" (Descartes 1642, 111). (See also Descartes's *Rules for the Direction of the Mind*, published posthumously in Latin in 1701, written around 1628.)

Note that neither Bacon nor Descartes was a New Atheist. Far from it! Both, however, wanted a world that was just matter in motion. This points to something really interesting. At the heart of what was going on in the Scientific Revolution, reflected in both the Renaissance and the Reformation, was a change in what linguists call "root metaphors." Until the beginning of the sixteenth century, the root metaphor was that of an organism (Ruse 2013). Plato, in the *Timaeus*, argued that the whole world is one big organism, created by the Demiurge, a facet apparently of the ultimate Form of the Good. Aristotle was more restrained. His Unmoved Mover does the only thing open to a perfect being, namely contemplating its own perfection. One trusts it had more reason than certain successful aspirants to the presidency of the

United States of America. The thing is that such a being is probably unaware of the physical world or its denizens, namely us. (We are aware of it, and our job is to strive for its perfection.) However, in Aristotelian philosophy, in the world itself, final cause rules supreme. Moreover, the place above all places where you see final cause in action is the organic world—the tree has roots in order to draw sustenance from the soil, and the eye serves the end of sight, something of great value to its possessor.

Platonic or Aristotelian, this kind of organicist thinking was very powerful and lasted right through the Middle Ages. Students of environmentalism have shown how miners digging for metals regarded the earth entirely as a mother who is giving sustenance to her children (Merchant 1980). She gives, and they must respect. Deniers of global warming would have been out of kilter as much back then as they are today. Then, as hinted above in talking of the Renaissance, it changed. Historian of science Rupert Hall put his finger firmly on the problem and the solution and what this meant:

No Christian could ultimately escape the implications of the fact that Aristotle's cosmos knew no Jehovah. Christianity taught him to see it as a divine artifact, rather than as a self-contained organism. The universe was subject to God's laws; its regularities and harmonies were divinely planned; its uniformity was a result of providential design. The ultimate mystery resided in God rather than in Nature, which could thus, by successive steps, be seen not as a self-sufficient Whole but as a divinely organized machine in which was transacted the unique drama of the Fall and Redemption. If an omnipresent God was all spirit, it was all the more easy to think of the physical universe as all matter; the intelligences, spirits, and Forms of Aristotle were first debased and then abandoned as unnecessary in a universe that contained nothing but God, human souls, and matter. (1954, xvi–xvii)

The root metaphor changed from that of an organism to that of a machine. It didn't happen overnight, and we find some embracing both metaphors. Johannes Kepler, of all people, a hero of the Revolution, was one. On the one hand, he enthusiastically embraced the machine metaphor: "It is my goal to show that the celestial machine is not some kind of divine being but rather like a clock." On the other hand, he was ever an enthusiastic Platonist, and the philosopher's Pythagorean fascination with numbers and geometrical shapes, not to mention the exalted state of the sun (the physical analogy to the Good), fed right into Kepler's science: "by the highest right we return to the sun, who alone appears, by virtue of his dignity and power, suited for this motive duty and worthy to become the home of God himself, not to say the first mover" (Burtt 1932, 48, quoting an early fragment). Heliocentrism may have put pressure on Christianity. The pagan Plato would have loved it. As also might Kepler's exploration in some detail of the analogies between the functioning of the earth's soul and more familiar bodily workings. The reader learns that "as the body displays tears, mucus, and earwax, and also in places lymph from pustules on the face, so the Earth displays amber and bitumen; as the bladder pours out urine, so the mountains pour out rivers; as the body produces excrement of sulphurous odor and farts which can even be set on fire, so the Earth produces sulphur, subterranean fires, thunder, and lightning; and as blood is generated in the veins of an animate being, and with it sweat, which is thrust outside the body, so in the veins of the Earth are generated metals and fossils, and rainy vapor" (Kepler 1619, 363–364). Sounds like the earth has been over-indulging in pease pudding.

Increasingly, however, the machine metaphor ruled supreme. The philosopher-chemist Robert Boyle is always good on these things. Using the same analogy as that of Kepler, he argued that the world is "like a rare clock, such as may be that at Strasbourg, where all things are so skillfully contrived that the engine being once set a-moving, all things proceed according to the artificer's first design, and the motions of the little statues that at such hours perform these or those motions do not require (like those of puppets) the peculiar interposing of the artificer or any intelligent agent employed by him, but perform their functions on particular occasions by virtue of the general and primitive contrivance of the whole engine" (1688, 12–13). Likewise, across the Channel, Descartes chipped in that we should think of our body "as a machine created by the hand of God, and in consequence incomparably better designed and with more admirable movements than any machine that can be invented by man" (Descartes 1642, 41).

Now note that this does not at once spell the death of God. At a metaphysical level, a machine is just as much into final cause—what came to be called teleology—as an organism. The vacuum is for cleaning the house (down the road) no less than an eye is for seeing (down the road). The crucial difference, and Hall sees this, is that the teleology of the organism is internal—often associated with Aristotle—meaning that it does not appeal outside the organism for value. The leaf is of value to the tree, whether or not humans exist, whether or not God exists. The teleology of the machine is external—often associated with Plato—in that it only gets its value from humans or God. A vacuum is just steel and plastic until someone fashions it and uses it. And this means the possibility is opened of sloughing off the designer. If you don't find final-cause talk very useful, then why bother at all with a Designer or Creator, at least in one's science? Not just spirits but God Himself gets kicked out. In the words of one of the most eminent historians of the Scientific Revolution, God became "a retired engineer" (Dijksterhuis 1961, 491).

The Problem of Organisms

At least in one's science. There was a very big fly in the ointment. No one was about to give up God, even though, as Isaac Newton shows, more and more people were moving to a kind of deism, God as Unmoved Mover and Jesus as just another chap—a very good chap but a chap nevertheless (Westfall 1980). What about organisms? They still seemed to demand final-cause thinking. This holds whether or not God exists, and if not God, then what? Boyle was fully aware of the dilemma. One of the organisms that he discussed in some detail was the bat, remarking: "Though bats be looked upon as a contemptible sort of creatures, yet I think they may afford us no contemptible argument to our present purpose" (1688, 194). He drew attention to the membranes between the digits that made the wings, the hooks on the wings for holding on to trees, the teeth for chewing, and in the case of the females, the internal organs for bringing forth live offspring and even the "dugs, to give such to her young ones" (196). The female is even restricted to two teats, because that is the number of offspring she has. Nevertheless, in seeking understanding, the scientist cannot speak to ultimate issues. This kind of inquiry is not open to the scientist qua scientist: "this is not the proper task of a naturalist, whose work, as he is such, is not so much to discover *why*, as *how*, particular effects are produc'd" (229–230). So, in today's language, as a scientist, Boyle pushed methodological naturalism—just laws, no design—but as a metaphysical non-naturalist, more than just methodological naturalism, he remained a Christian believer.

Politically and culturally, this sort of two-way thinking fit right into the English experience. Searching for a via media between the authority-insisting Roman Church and the biblical tyranny of the Calvinists, in the Elizabethan Compromise of the later sixteenth century, within the new Anglican Church, natural theology got inflated from its traditional theological status (Ruse 2003). A Boyle-type solution or amnesty served right through the eighteenth century, and much good (biological) understanding was achieved. There were tensions, especially as other aspects of what had happened in the sixteenth and seventeenth centuries were explored and extended. In the humanism realm, it wasn't just the ancient texts but modern works—especially by philosophers— that were continuing to gnaw away at traditional belief. David Hume is the prime example—although some of the French like Denis Diderot were not far behind (Ruse 2015). Hume's *Dialogues Concerning Natural Religion* contained a devastating attack on natural theology, especially the argument from design (the teleological argument) for God's existence. This matched other attacks, for instance, Hume's essay on miracles. Judged by the maxim "that no testimony is sufficient to establish a miracle, unless the testimony be of such a kind, that its falsehood would be more miraculous, than the fact, which it endeavors to establish" (Hume 1777, 115), it is simply not reasonable to believe in miracles rather than a naturalistic explanation like deliberate deceit. For all this, however, Hume ended as a deist, not a non-believer. Those organisms have to have some explanation, and natural explanations don't seem to do it.

> That the works of Nature bear a great analogy to the productions of art, is evident; and according to all the rules of good reasoning, we ought to infer, if we argue at all concerning them, that their causes have a proportional analogy. But as there are also considerable differences, we have reason to suppose a proportional difference in the causes; and in particular, ought to attribute a much higher degree of power and energy to the supreme cause, than any we have ever observed in mankind. Here then the existence of a DEITY is plainly ascertained by reason: and if we make it a question, whether, on account of these analogies, we can properly call him a mind or intelligence, notwithstanding the vast difference which may reasonably be supposed between him and human minds; what is this but a mere verbal controversy? (Hume 1779, 203–204)

Thus the legacy of the Renaissance. In the area of religion, likewise, we find the same chipping away and yet no real collapse. By the eighteenth century, voyages of discovery and then commerce were bringing Westerners into contact with quite alien religions, like

Buddhism and Hinduism. It was easy and natural just to regard these people as ignorant heathens, but greater and closer acquaintance with some of the beliefs rather shook Western complacency. *Unsophisticated* was just about the last term one would apply in many cases. Added to this was the rapid realization that nothing was worse for trade than religious strife. In India particularly, the British were very loath to allow Christian missionaries—it was only the pressure of evangelicals in the nineteenth century that let them in. However, it is all very well arguing that we have no right to impinge on already-established beliefs. Before long, people are going to start thinking that perhaps there is some value to them!

At the same time, scholars were turning their skills toward the Bible, wondering if it was quite the authentic record of God it was claimed to be. Did Moses really write those first five books? It was not until the first half of the nineteenth century that so-called Higher Criticism built up momentum, in Germany particularly, but even in the seventeenth century, the Dutch philosopher Spinoza was asking awkward questions. No wonder his fellow Jews excommunicated him. Always put all of this in perspective. At the end of the eighteenth century, we find Anglican divine Archdeacon William Paley churning out texts—*A View of the Evidences of Christianity* (1794) and *Natural Theology: or, Evidences of the Existence and Attributes of the Deity* (1802)—supporting both revealed and natural theology, but revealed theology really only because it measured up to natural theology. These were on the syllabus at Cambridge even into the twentieth century.

In the area of science, above all it was the great German philosopher Immanuel Kant who worried non-stop about final causes and their place in biology. You cannot get rid of them.

"An organized being is thus not a mere machine, for that has only a motive power, while the organized being possesses in itself a formative power, and indeed one that it communicates to the matter, which does not have it (it organizes the latter): thus it has a self-propagating formative power, which cannot be explained through the capacity for movement alone (that is, mechanism)" (Kant 1790, 246). And yet they are simply not acceptable in proper science. Uneasily, Kant concluded that they must have only a heuristic role, a point he conceded in a paradigmatically convoluted paragraph:

> The concept of a thing as in itself a natural end is therefore not a constitutive concept of the understanding or of reason, but it can still be a regulative concept for the reflecting power of judgment, for guiding research into objects of this kind and thinking over their highest ground in accordance with a remote analogy with our own causality in accordance with ends; not, of course, for the sake of knowledge of nature or of its original ground, but rather for the sake of the very same practical faculty of reason in us in analogy with which we consider the cause of that purposiveness. (1790, 247)

This didn't stop Kant from being rather nasty about biology, concluding that it would never live up to the physical sciences: "we can boldly say that it would be absurd for humans even to make such an attempt or to hope that there may yet arise a Newton who could make comprehensible even the generation of a blade of grass according to natural laws that no intention has ordered; rather, we must absolutely deny this insight to human beings" (271).

Charles Darwin

That, of course, was a challenge far too good to be ignored by an ambitious new researcher in the life sciences. A few decades later—in the 1830s—a Cambridge-educated young man by the name of Charles Darwin, fresh off a five-year voyage as ship's naturalist circumnavigating the globe on HMS *Beagle*, went right at the problem and showed that in biology, he rightfully could claim the mantle of Newton (Browne 1995; Ruse 2018b). Put him in context before introducing him. There were very few atheists at the beginning of the nineteenth century and not many agnostics. Deism of a stronger or weaker form was the general choice of those who had real doubts about the truth of Christianity. This was, for instance, the position of many of the revolutionaries in America. Ben Franklin: "I was scarce 15 when, after doubting by turns of several Points as I found them disputed in the different Books I read, I began to doubt of Revelation itself. Some Books against Deism fell into my Hands; they were said to be the Substance of Sermons preached at Boyle's Lectures. It happened that they wrought an Effect on me quite contrary to what was intended by them: For the Arguments of the Deists which were quoted to be refuted, appeared to me much Stronger than the Refutations. In short I soon became a thorough Deist" (c. 1790, 58).

The big problem was final cause. There was no convincing naturalistic explanation. What about evolution? In the *Critique of the Power of Judgment* (1790), Kant considered the possibility seriously; the analogies (homologies) between unrelated organisms—humans, apes, whales, even bats and birds—were strong evidence in its favor. Final cause was the block. Nevertheless, by the end of the eighteenth century, people were becoming evolutionists. Darwin's grandfather, Erasmus Darwin, was a well-known example. Although expectedly, there were many detractors, he had significant influence; it may well be the case that the German translation tipped the aged Kant. Although adaptations still could not be explained—the (false idea of the) inheritance of acquired characters (later called Lamarckism) went some way but by no means far enough—since organic evolution was seen as an epiphenomenon of ideas of social and cultural progress, that for most was reason enough. This did not mean that people jettisoned God. Rather, as is shown by the following verses of Erasmus Darwin—he was much given to poetry—there was continued movement in the direction of deism, a God who works through unbroken law. Here was a case where, apparently, science supported religion. Evolution is as much a support for deism, with its god of law, as it was later taken to be a refutation of theism, with its god of miracles.

> Organic Life beneath the shoreless waves
> Was born and nurs'd in Ocean's pearly caves;
> First forms minute, unseen by spheric glass,
> Move on the mud, or pierce the watery mass;
> These, as successive generations bloom,
> New *powers* acquire, and larger limbs assume;
> Whence countless groups of vegetation spring,
> And breathing realms of fin, and feet, and wing.
>
> Thus the tall Oak, the giant of the wood,
> Which bears Britannia's thunders on the flood;

> The Whale, unmeasured monster of the main,
> The lordly Lion, monarch of the plain,
> The Eagle soaring in the realms of air,
> Whose eye undazzled drinks the solar glare,
> Imperious man, who rules the bestial crowd,
> Of language, reason, and reflection proud,
> With brow erect who scorns this earthy sod,
> And styles himself the image of his God;
> Arose from rudiments of form and sense,
> An embryon point, or microscopic ens!
> (Darwin 1803, 1, V, 295–314)

Take note of how thoughts of progress, running up from the blob to the human, make the very backbone (to use an apt metaphor) of this vision. Much more on this later.

What was needed was for a professional scientist, probably British and soaked in natural theology, including—especially including—the central status of design, to get captivated by and converted to an evolutionary perspective. Such was Charles Darwin. On the *Beagle* voyage—from 1831 to 1836—he first lost his Christian faith, becoming a deist (because he could no longer believe in miracles), and then edged toward evolution, partly because of the fossil evidence but most strongly because of the idiosyncrasies of organic geological distributions. Why should the denizens of the Galapagos archipelago in the Pacific look so like the denizens of South America? Soon after his return to England—spring of 1837—Charles Darwin became an evolutionist, and then he searched frenetically for eighteen months to find a cause, the kind of equivalent of Newtonian gravitation. He soon spotted that selection might be the key. It was in the world of the farmyard or the bird or dog fancier, and he even came across its extension (in an 1809 pamphlet by a well-known breeder) into nature.

A severe winter, or a scarcity of food, by destroying the weak or unhealthy, has all the good effects of the most skilful selection. In cold and barren countries no animal can live to the age of maturity, but those who have strong constitutions; the weak and the unhealthy do not live to propagate their infirmities, as is too often the case with our domestic animals. To this I attribute the peculiar hardiness of the horses, cattle, and sheep, bred in mountainous countries, more than their having been inured to the severity of climate. (Sebright 1809)

Darwin took careful note of this passage, and even though he could not quite see the full import grasped that if something like this went on long enough, we would get full-blooded species. In a private notebook, he wrote of "excellent observations of sickly offspring being cut off so that not propagated by nature.—Whole art of making varieties may be inferred from facts stated" (Barrett et al. 1987, C 133).

I am not accusing Darwin of plagiarism. Sebright had no thoughts of evolution. Darwin did, and the problem was to make selection a universal and significant force for change. Then, in September 1838, Darwin read a work on political economy by the Reverend Thomas Robert Malthus, who—worried about potential population explosions (a real concern in Britain at the end of the eighteenth century when he first wrote)—argued that population potential always outstrips possible resources of food and space. Inevitably, there will be struggles for existence, and clearly even more for reproduction. Malthus was writing about humans, but since he acknowledged he was drawing on broader speculations by Benjamin Franklin, Darwin had little problem in generalizing back again and going from the struggle to its consequences. Nigh uniquely in the history of science, we can document the very moment of insight. This is from a private journal (which explains its somewhat staccato style):

> Even a few years plenty, makes population in Men increase & an ordinary crop causes a dearth. take Europe on an average every species must have same number killed year with year by hawks, by cold &c.—even one species of hawk decreasing in number must affect instantaneously all the rest.—The final cause of all this wedging, must be to sort out proper structure, & adapt it to changes.—to do that for form, which Malthus shows is the final effect (by means however of volition) of this populousness on the energy of man. One may say there is a force like a hundred thousand wedges trying force into every kind of adapted structure into the gaps of in the oeconomy of nature, or rather forming gaps by thrusting out weaker ones. (in Barrett et al. 1987, D 135)

Malthus (1826) had written: "It may safely be pronounced, therefore, that the population, when unchecked, goes on doubling itself every twenty five years, or increases in a geometrical ratio."

Darwin had gotten his cause. What he had to do now was to fold it all into a coherent evolutionary theory, which he did over the next five years or so. Then he sat on his ideas for the next years until nigh the end of the 1850s. More later for possible reasons why. Finally, spurred by the arrival of an essay from a younger naturalist—Alfred Russel Wallace—Darwin was moved to action, and *On the Origin of Species by Means of Natural Selection, or the Preservation of Favoured Races in the Struggle for Life* appeared at the end of 1859. Darwin's style is casual, but the theory is carefully and professionally constructed (Ruse 1979a; Ruse 2008; Richards and Ruse 2016). The aim was to show that natural selection, like Newtonian gravitational force, is what the philosophers called a *vera causa*, a true cause. There is some ambiguity about the precise meaning of this term, with some of an empiricist bent, like the physicist John F. W. Herschel (1830), insisting that one had to find analogies with everyday experience. Others of a rationalist bent, like the scientist and then philosopher and historian of science William Whewell (1840), insisted that one had to show the cause explained in many different areas. This form of argument was what Whewell called a "consilience of inductions." No need for direct experience.

Darwin covered both options! He opened with a discussion of human selection of organisms for our ends—farm animals and crops, from sheep to turnips, and organisms shaped by us for pleasure, like birds and dogs. He made much of the fact that pigeons have such variety and yet clearly all come from common stock. Then, having postulated that new variations are constantly appearing in populations—not uncaused but undirected—he was ready for his key inferences. These were put in (quasi- or proto-) deductive form. As the philosophers pointed out constantly, this was the form of the gravitational theory of Newton. First, the Malthusian element to a struggle for existence:

A struggle for existence inevitably follows from the high rate at which all organic beings tend to increase. Every being, which during its natural lifetime produces several eggs or seeds, must suffer destruction during some period of its life, and during some season or occasional year, otherwise, on the principle of geometrical increase, its numbers would quickly become so inordinately great that no country could support the product. Hence, as more individuals are produced than can possibly survive, there must in every case be a struggle for existence, either one individual with another of the same species, or with the individuals of distinct species, or with the physical conditions of life. (Darwin 1859, 63–64)

Then, second, to natural selection:

Let it be borne in mind how infinitely complex and close-fitting are the mutual relations of all organic beings to each other and to their physical conditions of life. Can it, then, be thought improbable, seeing that variations useful to man have undoubtedly occurred, that other variations useful in some way to each being in the great and complex battle of life, should sometimes occur in the course of thousands of generations? If such do occur, can we doubt (remembering that many more individuals are born than can possibly survive) that individuals having any advantage, however slight, over others, would have the best chance of surviving and of procreating their kind? On the other hand, we may feel sure that any variation in the least degree injurious would be rigidly destroyed. This preservation of favourable variations and the rejection of injurious variations, I call Natural Selection. (80–81)

The all-important point is that this cause—natural selection—points to adaptation, to final cause. The eye is created as it is in order to see. The flower to attract pollinators. The fangs of the snake to kill. The instincts of the nest-building bird to promote and continue life. It is not just change but change of a particular cause:

> Under nature, the slightest difference of structure or consti-tution may well turn the nicely-balanced scale in the struggle for life, and so be preserved. How fleeting are the wishes and efforts of man! how short his time! and consequently how poor will his products be, compared with those accumulated by nature during whole geological periods. Can we wonder, then, that nature's productions should be far "truer" in char-acter than man's productions; that they should be infinitely better adapted to the most complex conditions of life, and should plainly bear the stamp of far higher workmanship? (83–84)

"Of far higher workmanship"?! Strong echoes here of Darwin having had Paley on design drilled into him at Cambridge. Natural selection wasn't just any old mechanism.

Although notice, as one always must in discussing important moves in Western intellectual culture, the underlying dynamics of Plato versus Aristotle. We can think of the Copernican Revolution as the triumph of Plato over Aristotle—moving the sun to the center with the added unforeseen consequence that God now being an external designer—as Plato stressed in the *Timaeus*—can now be pushed out of science. Equally, we can think of the Darwinian Revolution as the triumph of Aristotle over Plato. Darwin learnt his final-cause thinking from the Platonist Paley, but now the teleology is going to be incorporated into the science without need of God-talk, in a way that Aristotle would have thought obligatory. The point is that conscious intention is not part of the story. "This is most obvious in the animals other than man: they make things neither nor after inquiry or deliberation . . . If then it is both by nature and for an end that the swallow makes its nest and the spider its web, and plants grow leaves for the sake of the fruit and send their roots down (not up) for the sake of nourishment, it is plain that this kind of cause is operative in things which come to be and are by nature" (Aristotle, *Physics*, 1984, 340, 20–29).

Now, having first argued that this final-cause force of natural selection could lead to the tree of life, Darwin moved into the final phase of his theory, as he argued that he could explain phenomena right across the life sciences. Social behavior, the fossil record (paleontology), geographical distributions (biogeography), anatomy and morphology, systematics, embryology, vestigial organs. Why is there a progressive fossil record? Evolution through natural selection. Why the Linnaean system? Because it reflects the tree of life. Why are there such similarities between the organisms of very different adult forms? Because selection only works on the adults. This all done, Darwin was ready for his final famous passage:

> It is interesting to contemplate an entangled bank, clothed with many plants of many kinds, with birds singing on the bushes, with various insects flitting about, and with worms

crawling through the damp earth, and to reflect that these elaborately constructed forms, so different from each other, and dependent on each other in so complex a manner, have all been produced by laws acting around us.... Thus, from the war of nature, from famine and death, the most exalted object which we are capable of conceiving, namely, the production of the higher animals, directly follows. There is grandeur in this view of life, with its several powers, having been originally breathed into a few forms or into one; and that, whilst this planet has gone cycling on according to the fixed law of gravity, from so simple a beginning endless forms most beautiful and most wonderful have been, and are being, evolved. (489–490)

In the *Origin*, Darwin said virtually nothing about our own species, *Homo sapiens*. We know that he always included us in the story. Indeed, in his private notebooks, the first clue that we have that Darwin is now working with natural selection is an application to humans, and not just to our bodies but to our mental abilities. In his *Autobiography*, written toward the end of his life, Darwin reaffirmed this belief: "I am inclined to agree with Francis Galton [Darwin's half cousin] in believing that education and environment produce only a small effect on the mind of any one, and that most of our qualities are innate" (Darwin [1887] 1958, 43). From the first to the last, Darwin was (in today's terms) a human sociobiologist or evolutionary psychologist. In the *Origin*, however, Darwin wanted to get his main theory into the public domain. So, not to appear cowardly, he simply said that "light will be thrown on man and his history" (1859, 488). No one was fooled, and at once Darwin became known as the author of the "monkey theory" or the "gorilla theory."

Probably, if he had been a free agent, he would have stayed on the sidelines of the human evolution debate—Darwin was not fond of controversy—but in the 1860s, Wallace turned to spiritualism, arguing that only supernatural forces could explain human evolution. Horrified, Darwin again put pen to paper, and in 1871, he published *The Descent of Man, and Selection in Relation to Sex*. Much of the early part of the book is predictable, as Darwin argued that his theory applies to us. He spent considerable time on the evolution of morality, arguing that cooperation is a major positive adaptation for human beings:

> It must not be forgotten that although a high standard of morality gives but a slight or no advantage to each individual man and his children over the other men of the same tribe, yet that an advancement in the standard of morality and an increase in the number of well-endowed men will certainly give an immense advantage to one tribe over another. There can be no doubt that a tribe including many members who, from possessing in a high degree the spirit of patriotism, fidelity, obedience, courage, and sympathy, were always ready to give aid to each other and to sacrifice themselves for the common good, would be victorious over most other tribes; and this would be natural selection. At all times throughout the world tribes have supplanted other tribes; and as morality is one element in their success, the standard of morality and the number of well-endowed men will thus everywhere tend to rise and increase. (Darwin 1871, 1, 166)

Darwin always thought of tribes as interrelated individuals (or those who think of themselves as interrelated), so from a reproduction viewpoint it was always a matter of helping others and thereby helping yourself (Richards and Ruse 2016). In the language of Richard Dawkins (1976), though Darwin had no true grasp of heredity, it is "selfish genes" all the way. Even if you don't reproduce yourself, your blood relatives will. Darwin drew on the analogy of the farmyard, where although the steer does not reproduce, his bloodline does, inasmuch as he has desired features and is worth repeating.

What makes the *Descent* rather odd is that most of it is not about humans at all! It is a very detailed and lengthy discussion of Darwin's second mechanism of "sexual selection," where the competition is within a species for mates—such as the peacock having a ludicrously large and flamboyant tail to attract the peahen. The reason for the imbalance is simple. Wallace argued that certain human features, like hairlessness, could not have been produced by natural selection and hence demanded spirit forces. Darwin agreed that natural selection could not do the job but argued that sexual selection could! Hairlessness was all a matter of people choosing those mates they found most attractive. All of which led to some very Victorian sentiments:

Man is more courageous, pugnacious, and energetic than woman, and has a more inventive genius. His brain is absolutely larger, but whether relatively to the larger size of his body, in comparison with that of woman, has not, I believe been fully ascertained.

Continuing:

> Male and female children resemble each other closely, like
> the young of so many other animals in which the adult sexes
> differ; they likewise resemble the mature female much more
> closely, than the mature male. The female, however, ulti-
> mately assumes certain distinctive characters, and in the for-
> mation of her skull, is said to be intermediate between the
> child and the man. (Darwin 1871, 2, 317)

Darwin was a great revolutionary. He was no rebel against the
mores and norms of Victorian society.

Evolution after Darwin

What of the aftermath? Let us deal first with the science and
turn then to religion and the quest for meaning. A popular account of the fate of Darwin's theory has an almost immediate
conversion of scientists to evolution and an equally great re-
luctance to accept natural selection (Bowler 1988, 2013). This
mechanism only came into its own in the twentieth century,
when it had the backing of Mendelian—later molecular—genetics. Even now, there are serious concerns. Well, the first part
of this story is true, at least. Almost overnight, like the emperor's
new clothes, when Darwin said it evolves, people said they had
known that all along. The second part of the story really isn't
true, although some things—other than a desperate religious

or quasi-religious desire to avoid embracing natural selection—do give a cover for the myth. It is true that for the rest of the nineteenth century, many active biologists downgraded selection, mainly because it didn't speak to the kinds of issues with which they were concerned. No one could see selection at work on the dinosaurs, for instance. When you are doing systematics, as Darwin himself discovered when doing a massive study of the barnacles, adaptation gets in the way! You are looking for homologies to discern underlying relationships. However, expectedly in the areas where selection could help—especially when dealing with fast-breeding organisms like insects—it was prized (Kimler and Ruse 2013; Ruse 2017a; Ruse 2018b). Most notably, shortly after the *Origin* appeared, a sometime traveling companion of Wallace, Henry Walter Bates, proposed a selection-based theory of mimicry, arguing that some butterflies and moths—non-poisonous—mimic other butterflies and moths—poisonous. The biggest predators are birds, and they learn to avoid the poisonous forms. The non-poisonous forms slip in under cover, as it were. They have to be uncommon—as they are—otherwise the birds would ferret out the deceit and all would fail. A few years later, a German-Argentinian naturalist, Fritz Müller, found another form of mimicry, which today bears his name, as Bates's discovery bears his name. Much insightful selection-based work was done on industrial melanism—where insects became darker to be better camouflaged against trees that, thanks to the Industrial Revolution, were already becoming darker. One enthusiast even wrote to Darwin about this:

My dear Sir,

The belief that I am about to relate something which may be of interest to you, must be my excuse for troubling you with a letter.

Perhaps among the whole of the British Lepidoptera, no species varies more, according to the locality in which it is found, than does that Geometer, Gnophos obscurata. They are almost black on the New Forest peat; grey on limestone; almost white on the chalk near Lewes; and brown on clay, and on the red soil of Herefordshire.

Do these variations point to the "survival of the fittest"? I think so. It was, therefore, with some surprise that I took specimens as dark as any of those in the New Forest on a chalk slope; and I have pondered for a solution. Can this be it?

It is a curious fact, in connexion with these dark specimens, that for the last quarter of a century the chalk slope, on which they occur, has been swept by volumes of black smoke from some lime-kilns situated at the bottom: the herbage, although growing luxuriantly, is blackened by it.

I am told, too, that the very light specimens are now much less common at Lewes than formerly, and that, for some few years, lime-kilns have been in use there.

These are the facts I desire to bring to your notice.

I am, Dear Sir, Yours very faithfully,

A. B. Farn

(Letter from Albert Brydges Farn, November 18, 1878, Darwin Correspondence Project, 11747)

When Mendelism was discovered at the beginning of the new century, it was indeed extended and developed (Provine 1971;

Ruse 1996). This was thanks particularly to Thomas Hunt Morgan and his students at Columbia University, who in the second decade of the century produced the "classical theory of the gene." This, as we still believe today, located the units of heredity in real time and space on the chromosomes in the nuclei of cells. The theory was generalized to populations—random variation (mutation) being acted upon by selection—and so was the beginning of modern evolutionary thinking, labeled neo-Darwinism (or just Darwinism) in Britain and the synthetic theory of evolution in America. In Britain, the theoreticians were Ronald Fisher (1930) and J. B. S. Haldane (1932); in America, Sewall Wright (1931; 1932). Then came the empiricists, notably E. B. Ford (1964) and students in England and the Russian-born Theodosius Dobzhansky (1937) in America, soon joined by the German-born systematist Ernst Mayr (1942), the paleontologist George Gaylord Simpson (1944), and at mid-century the botanist G. Ledyard Stebbins (1950). In Kuhnian language, there was a fully functioning paradigm, with natural selection at its core. It was soon molecularized, but essentially this was more of the same, rather than a new revolution overthrowing the old (Ruse 1973).

What of natural selection? No one ever claimed it was the only force for change. Darwin was ever a Lamarckian, although now, with the coming of Mendelism, that was long gone. Fisher trained as a physicist, and he always regarded populations like gases, with molecules/genes dashing around in the cloud and selection (like pressure) working on the whole. Wright worked for years with the US Department of Agriculture on breeding shorthorn cattle. He tended to think in terms of small, isolated groups that change and then come back into the big population with new variations to spread. The interesting thing about small populations is that chance can override the power of selection, leading to the production of features not immediately adaptive but possibly with great potential. For a short while, this idea of "genetic drift" was very trendy. Dobzhansky made much use of it in the first edition of his groundbreaking *Genetics and the Origin of Species* (1937). But soon it was found—by Dobzhansky, of all people—to be not nearly as common as he supposed. Today, it has a much-reduced profile. Except at the molecular level. Down there, drift is thought important because change can escape the forces of selection. This realization has led to sophisticated methods of estimating absolute evolutionary times—when a population splits, for instance (Ayala 2009).

"It is a truth universally acknowledged, that a single man in possession of a good fortune, must be in want of a wife." It is likewise a truth universally acknowledged that natural selection continues to get up people's noses. At heart it is seen as unfeeling, reductionistic, or some such thing, rather than holistic and making for universal brotherhood (Reiss and Ruse forthcoming). Metaphors like "selfish genes" don't much help. Always, there are critics looking for a warmer, fuzzier causal method of change. Today, evolutionary development—"evo-devo"—is the big thing, and again, with boring inevitability, it is argued that selection is under threat. What no one seems to realize is that any good theory is going to go on developing—that's what makes science both fun and challenging. Neo-Darwinism did rather treat organisms as black boxes—not entirely—and molecular biology has liberated us. Nevertheless, it is now claimed by critics that "epigenetics" shows Darwinism to be false or redundant, because in a kind of neo-Lamarckian fashion, changes of or pressures on the organism can be inherited. The truth is rather less dramatic. There are perfectly good molecular explanations of this, it happens rarely, and it is never long-lasting—just a generation or two. The eye was still made by natural selection. Believe me, folks! (Actually, don't believe me. Jerry Coyne's blog, "Why Evolution Is True," is as good on science as it is silly on philosophy. Do a search for "epigenetics," find out about the whole subject for yourself, and then make up your own mind.)

The Loss of Meaning

"What is Darwinism?" asked Charles Hodge (1874), principal of Princeton Theological Seminary and leading Calvinist theologian of the mid-nineteenth century. His conclusion: It is atheism. This is not true, actually. At least, it is not true in the eyes of many Christians and more than a few non-believers, including myself. Some people of faith, indeed, have embraced it with enthusiasm. A good example is the late-nineteenth-century High Anglican Aubrey Moore. He wrote:

> Science had pushed the deist's God farther and farther away, and at the moment when it seemed as if He would be thrust out altogether, Darwinism appeared, and, under the guise of a foe, did the work of a friend. It has conferred upon philosophy and religion an inestimable benefit, by showing us that we must choose between two alternatives. Either God is

everywhere present in nature, or He is nowhere. He cannot
be here, and not there. He cannot delegate his power to
demigods called "second causes." In nature everything must
be His work or nothing. We must frankly return to the
Christian view of direct Divine agency, the immanence of
Divine power from end to end, the belief in a God in Whom
not only we, but all things have their being, or we must banish
him altogether. (1890, 99–100)

For now, I am going to leave this discussion at the philosophical/
theological level. This will be an opening topic of the next chapter.
Here, I want to follow up the point made above that both believers
and non-believers embraced evolutionary thinking, Darwinian
thinking even. This rather suggests, which is true, that the main
reason driving people to non-belief was not Darwinism as such
(Benn 1906). Biography after biography of nineteenth-century
figures tell us that the real reasons for leaving Christianity were
almost always theological (Budd 1977).

On the one hand, there was the corroding force of German
Higher Criticism, arguing that the Bible is just a collection of fallible old human-inspired and written documents. Stories like the
Resurrection were not literally true but were fairy tales concocted
to meet eschatological expectations. At the same time, there was a
feeling that some of the events and interpretations were unacceptable to the point of total rejection. To think that Jesus turned a few
loaves and fishes into food for thousands is to turn him into an em-
ployee of a grocery store—Whole Foods in America or Sainsbury's
in England. Even worse were claims about substitutionary atonement. The very idea that someone has to suffer dreadfully to buy

off God is in itself repellent. To think that Jesus could do it on my behalf is simply grotesque. No morality here.

On the other hand, people like Hume were having an effect. Natural theology was falling out of favor. The Danish theologian Søren Kierkegaard (1944) argued that Christianity demands a leap of faith, and proving the existence of God rather undercuts this. The greatest British theologian of the nineteenth century, John Henry Newman, whose spiritual path took him from an Anglican evangelical childhood, through the very High Church—the Oxford Movement—and on over to Rome, where he ended as a cardinal and prince of the Church, was very wary of design arguments. He wrote: "I believe in design because I believe in God; not in a God because I see design." He continued: "Design teaches me power, skill and goodness—not sanctity, not mercy, not a future judgment, which three are of the essence of religion" (Newman 1973, 97).

Darwin was typical. On the *Beagle* voyage, because he found he could not believe in the biblical miracles, he lost his belief in Jesus as Savior. He continued as a deist right through the writing of the *Origin*, finally fading to agnosticism in the 1870s. He was explicit. What really upset him was the thought that non-believers would go straight to hell: "I can indeed hardly see how anyone ought to wish Christianity to be true; for if so the plain language of the text seems to show that the men who do not believe, and this would include my Father, Brother and almost all my best friends, will be everlastingly punished. And this is a damnable doctrine" (1887, 87).

If the theory of evolution through natural selection had a more nuanced relationship to Christianity, what indeed was this relationship? It was something as crucial to our discussion as it is possible for a topic to be. It undercut the belief in a loving and caring God. It may not have made atheism mandatory, but with the naturalistic explanation of final cause, it made non-belief possible. In the immortal words of Richard Dawkins (1986), it was now possible to be an "intellectually fulfilled atheist." Moreover, God, even if He does exist, seems now to be totally indifferent to our well-being and status. Thanks to the struggle for existence, life truly is a Hobbesian nightmare. Far from creating us in His image, God didn't do it and simply couldn't care less about us. Like Aristotle's Unmoved Mover, He may not even know of us. Unlike Aristotle's Unmoved Mover, there is no reason to think Him perfect and no reason to strive to emulate Him. An early poem from 1866 by Thomas Hardy, raised an Anglican but bereft of hope after the *Origin*, puts it bluntly:

> If but some vengeful god would call to me
> From up the sky, and laugh: "Thou suffering thing,
> Know that thy sorrow is my ecstasy,
> That thy love's loss is my hate's profiting!"
>
> Then would I bear it, clench myself, and die,
> Steeled by the sense of ire unmerited;
> Half-eased in that a Powerfuller than I
> Had willed and meted me the tears I shed.
>
> But not so. How arrives it joy lies slain,
> And why unblooms the best hope ever sown?
> —Crass Casualty obstructs the sun and rain,
> And dicing Time for gladness casts a moan. . . .

> These purblind Doomsters had as readily strown
> Blisses about my pilgrimage as pain.
> ("Hap," 1866, in Hardy 1994, 5)

Crass Casualty, dicing Time, purblind Doomsters! It is not that God permits evil and even occasions it, as one might take the message of Job, but that God is as likely to be friendly as hurtful. In the world of the struggle for existence and natural selection, everything, including us humans, is simply the product of the blind forces of nature—no rhyme, no reason, no meaning or Meaning. Forget about eternal bliss and that sort of thing. You are not going to get it down here, and you are not going to get it up there. Truly, in the words of Camus, nigh a century later, life is absurd.

It is this aspect of Darwinism—note, not just evolution— that makes it so all-important in our discussion. Life has no Meaning. Get over it. Or need we? Is this too hasty a conclusion? In the next two chapters, we will look at two responses— one from the religious and one from the non-religious, from Darwinians, of all people—who argue that there is still Meaning to what we all do. Meaning that is objective, given to us, not subjective, created by us.

2 | HAS RELIGION REALLY LOST THE ANSWER?

Are Science and Religion at War?

Steven Weinberg, Nobel Prize-winning physicist, wrote: "The more the universe seems comprehensible, the more it also seems pointless" (1997, 154). He is part of a tradition going back at least to the nineteenth century of people—mainly scientists—who argue that science and religion are necessarily opposed. They are at war, and science wins on all fronts. Religion is out-of-date superstition. Take man-of-the-blog Jerry Coyne, who—for all my teasing—truly is a brilliant scientist. The very title of his latest book tells all: *Faith versus Fact: Why Science and Religion Are Incompatible* (2015). Endorsed on the cover by none other than Richard Dawkins: "It's hard to see how any reasonable person can resist the conclusions of his superbly argued book." Coyne describes Darwin's *Origin* as the "greatest Scripture-killer ever penned," claiming that "the book demolished (not deliberately) an entire series of biblical claims by demonstrating that purely naturalistic processes—evolution and natural selection—could explain patterns in nature previously explainable only by invoking a Great Designer" (2). This chapter's title question is answered already. Move on to the next chapter.

As you might expect, not everyone—not only Christians and other believers—is quite so sure that the matter is so definitively settled (Barbour 1990; Ruse 2010). The only time in my life I have been inspired to write anything of even a semi-fictional nature, I penned a short dialogue with a Coyne-like figure facing no fewer than four opponents (Ruse 2016). Naturally, these included a wise and learned philosopher, a disinterested seeker of the truth, whose name—Martin Rudge, drawn from characters in Dickens— coincidentally has the same initials as mine. For a start, as all these critics pointed out, there is the assumption that when science and religion—faith and fact, to use Coyne's language—go head to head against each other, naked mud-wrestling, as one might say, whatever science says trumps whatever religion says. Science says there was no universal flood. Religion says there was. Science wins, because science is science and not religion.

There are clearly some Christians who would dispute this. The Calvinist philosopher Alvin Plantinga (2000b) is one. He claims to have a *sensus divinitatis*, which, as it were, gives him direct Skyping privileges with God. Faith tells him something, and that is that. However, I don't think we need to worry too much here, and I suspect that even Plantinga—if not all others—would find some way of arguing that there is no real conflict. Truth cannot be opposed to truth. We saw in chapter 1 how after the Reformation there was a sense that doing science was God's work, and that is surely the position of most Christians today. Science in itself is not anti-God. Far from it. We are using our God-given powers of sense and reason—made as we are in His image—to discover and understand His creation. And in a tradition back to and beyond Augustine, there are warnings about taking the Bible as science. God could not tell the rather naive Israelites about the world in the language of science and mathematics. They would have been completely out of their depth. They were not sophisticated Romans like us.

That said, there is no question that parts of the Bible do conflict with science. Not all of this is thanks to Darwin. The Bible claims that Noah's Flood was universal. It covered the whole earth. Back in the 1820s, the professor of geology at the University of Oxford, William Buckland, thought he had found proof of the Flood in Yorkshire in the north of England (Ruse 1979a). There were fossil relics in a cave that he claimed pointed to a massive deluge as having killed off all the animals. He was ridiculed, including by the professor of geology at Cambridge, Adam Sedgwick—the man who was to write scathing letters to Darwin after the *Origin*—and that was that. Science and religion can and do conflict. Although this is a nice example where many, including myself, would argue that conflict is more apparent than real. Think about the Noah story. People are misbehaving, and so God decides to wipe them all out. "And the Lord said, I will destroy man whom I have created from the face of the earth; both man, and beast, and the creeping thing, and the fowls of the air; for it repenteth me that I have made them" (Genesis 6:7). The exceptions are Noah and his family, who are to seed the earth all over again. And what happens after the Flood? After you leave primary school, you learn that Noah takes off all his clothes and gets blind drunk—sounds like quite a party—and his son takes a gander and tells everyone. "And he drank of the wine, and was drunken; and he was uncovered within his tent. And Ham, the father of Canaan, saw the nakedness of his father, and told his two brethren without" (9:21–22). Noah and the Ark is not a story about what happens when you have too much rain (Pleins 2003). For my money, it is a story that tells you that simple solutions to complex problems just don't work. Wipe out evil at one stroke. Surprise! Surprise! Evil is still around. Would that those good Christians, George W. Bush and Tony Blair, had taken note of the story before they marched into Iraq.

However, there are some parts of science that really do cause problems for religion: claims of Darwinian evolutionary biology that cause problems for Christianity. Remember, in the West, the standard explanation of the need for Jesus and his death on the Cross—the explanation of both Catholics and those especially in the Lutheran and Calvinist traditions (and which is in the Anglican Thirty-Nine Articles)—is due to Saint Augustine. Adam fell, and this brought on original sin. We, his descendants, are all tainted. We may not be born sinners, but we have inclinations that way. Original sin "is the fault and corruption of the Nature of every man, that naturally is engendered of the offspring of Adam; whereby man is very far gone from original righteousness, and is of his own nature inclined to evil, so that the flesh lusteth always contrary to the Spirit; and therefore in every person born into this world, it deserveth God's wrath and damnation" (Article 9). The only way we could be saved is through a sacrifice, and only a sacrifice of God Himself would do the trick. Substitutionary atonement! God on the Cross to save us. But if Darwinism is true, this must be false. We know that human evolution went through some bottlenecks, but at no point were we much below several thousand. There was no unique original pair, nor was there a time when humans were sinless. They were always like the way we are now, sometimes nice and sometimes not. So were their moms and dads and uncles and aunts, all the way back to the apes. And if Jane Goodall is right, they weren't always such great pals, either. Chimpanzees go out on raiding parties, intending to kill other chimpanzees (Ruse 2018b).

One should say at once that from the start, not everyone has bought into the Augustinian story. Eastern Christianity never did. When I first heard it, as one raised a Quaker, I simply could not believe it. Substitutionary atonement is so pagan. So repellent to any decent-minded person. That someone should need to die in agony for my sins two thousand years later is ethically disgusting and totally ludicrous. To be fair, my acquaintance with pagans made it clear that today they, too, have little time for this kind of religious sadism. The most interesting character in my little dialogue—based on a couple of truly remarkable (and rather wonderful) people I met when researching a book on the Gaia hypothesis (Ruse 2013)—was Peaseblossom, a transsexual pagan from California. As she made very clear, the only sacrifices pagans are into are each other's virginity. Fortunately, there are alternatives—to substitutionary atonement, that is. Another tradition goes back earlier than Augustine to Irenaeus of Lyons, who in his work *Against Heresies* sees us made developmentally as beings that must grow and escape from sin. Jesus is an example of perfect love, not a blood sacrifice. "Our Lord Jesus Christ, who did, through His transcendent love, become what we are, that He might bring us to be even what He is Himself" (c. 180, 368). Or as Friedrich Nietzsche, of all people, put it: "This 'bringer of glad tidings' died as he had lived, as he taught—not to 'redeem mankind' but to demonstrate how one ought to live" (1895, 159). Rather than the agony of the wracked figure in a Grünewald Crucifixion, we have the loving friend of Mary and Martha in Vermeer's great early painting.

Questions beyond Science?

In the light of modern science, there are ways to save Christianity. Do not underestimate how major some of the changes might have to be. But isn't the whole enterprise ultimately doomed to failure? Adam and Eve today, Jesus tomorrow? After all, who needs a Creator God when we have the Big Bang? Here, the Christians—let us stay with them for our discussion—would argue that there are some questions that are simply beyond science. They would argue that such questions always will be beyond science. They would argue that religion can rightly step in and answer such questions. In fact, not only Christians argue this. Stephen Jay Gould (1999) famously—notoriously in the eyes of some—argued for what is known somewhat disparagingly as "accommodationism." There will always be questions beyond science for religion to answer. He spoke of twin "Magisteria," meaning twin perspectives on viewing the world, and he claimed that these do not compete—hence, science and religion are independent, not conflicting. Properly understood, they cannot conflict. Unfortunately, however, Gould's idea of the Magisterium for religion confined it entirely to ethical questions. Science can tell us about the world. Religion can tell us, and only tell us, about how we should behave in the world. Science tells us that males are sexual aggressors. Religion/ethics tells us that males should restrain themselves. Unfortunate, because the simple fact is that religion, the Christian religion, wants to go way beyond this. It talks of a Creator God, it talks of Jesus rising from the dead, it talks of a life hereafter. These are ontological claims, not ethical claims.

I don't want to get too bogged down here in the Jesus story. If He really did rise from the dead, then it has a lot going for it. If He did not, then the believer still has a time-honored method of reply. Miracles should be judged by meaning, not whether they violate the laws of nature (Ruse 2001a). Take the Resurrection. The eminent philosopher of religion John Hick wrote of his conversion to Christianity as a late teenager. He was resisting the call. Then it happened. "An experience of this kind which I cannot forget, even though it happened forty-two years ago [1942], occurred— of all places—on the top deck of a bus in the middle of the city of Hull.... As everyone will be very conscious who can themselves remember such a moment, all descriptions are inadequate. But it was as though the skies opened up and light poured down and filled me with a sense of overflowing joy, in response to an immense transcendent goodness and love" (Hick 2005). Was this psychological? Was it direct intervention? Who cares? It happened, and it mattered. The same for the disciples on the Third Day. They were despondent. Their lord and master was dead, killed as a common criminal. Then, suddenly, they knew He lived, and all made sense. Was this group hysteria, self-deception, or a direct divine intervention? Who cares? It happened, and it mattered.

I am giving a pass to a lot, and I expect already many readers will be disagreeing. Let's move on and turn to the big questions about

Christianity, especially God as Creator and (Weinberg's bugaboo) consequent claims about the world having meaning. Although I disagree with Gould's conclusion about religion as morality only, he is on the right track. Religion does—and must—speak to that which is independent of science (Ruse 2010). How is this possible? In his incredibly long and self-indulgent book *A Secular Age*, Canadian philosopher Charles Taylor spots what is going on. He puts the matter in terms of materialism rather than naturalistic science, but the point is made. Materialism simply doesn't address some questions, like "thinking and conceptual spontaneity" and "motivation and the validity of ethics" (2007, 597). The question, which he leaves unanswered, is why? Following Thomas Kuhn (1962), who later in life argued that paradigms in some sense are metaphors, I argue that here is the clue. Metaphors are invaluable in science, but, as Kuhn argued about paradigms, part of their success is—to use another metaphor—that as in horse racing, they put blinkers on the scientist (Kuhn 1993). They direct the scientist to important soluble problems and deflect him or her from the unimportant or insoluble. Take an example from poetry, and modify Robert Burns from a simile to a metaphor: "My love is a red, red rose." We know what the poet means. His love is beautiful in a radiant sort of way. She is young and fresh, not raddled by years and experience. If he is joking, he might be referring to the fact that his love tends to be a bit prickly and should be handled with care. Is his love good at mathematics? Is his love a Protestant or a Catholic? As it happens, these are answerable questions, but not in this context. They are irrelevant.

Same in science. Darwin makes great use of the metaphor, taken ultimately from Adam Smith, of a "division of labor." As in factories, organisms succeed by specializing. This, however, takes them further and further apart, until they are irretrievably different. He uses the concept to explain the caste system in social insects: "As ants work by inherited instincts and by inherited tools or weapons, and not by acquired knowledge and manufactured instruments, a perfect division of labour could be effected with them only by the workers being sterile; for had they been fertile, they would have intercrossed, and their instincts and structure would have become blended. And nature has, as I believe, effected this admirable division of labour in the communities of ants, by the means of natural selection" (Darwin 1859, 242).

Thanks to the metaphor, Darwin doesn't waste time thinking about on which continent these insects are found. Perhaps later he could show this to be relevant, but for now, heads down and look at the nests.

Meaning

Take now the root metaphor of modern science, the world as a machine. Incredibly fertile and mind-focusing. But there are questions it doesn't even try to answer, pertinent to our discussion because they speak directly to the problem of m/Meaning. Here is a list of four. Like William Whewell (1840), I am a bit of a historicist about my philosophical thinking. I don't look upon the list as written in stone. There may be more questions. Some, I doubt all, will start to look less absolute. You will see in chapter 3 how I am starting to worry a little—just a little!—about one of them. This said, the first is what Martin Heidegger (1959) calls the "fundamental question" of metaphysics: Why is there something rather than nothing? This does not mean something temporal. Contrary to a former pope and some prominent physics popularizers (Krauss 2012), the Big Bang is not an answer. The question is about the very nature of existence. Why the Big Bang? Some philosophers, notably Ludwig Wittgenstein (1965), argue that there can be no solution, so it is not a genuine question. I don't think you can simply declare something off limits like this. The fact that we cannot answer it doesn't mean it is not a genuine question. Has the United States of America reached the tipping point, and is it about to decline and fall like the empires of Rome and Britain before it? I cannot answer this, but it is a genuine question. In the case of Heidegger's question—Why is there something rather than nothing?—the machine metaphor in which modern science is embedded shows why it is not a scientific question. The metaphor takes existence and origins for granted. Of course, you can ask where Ford got the aluminum it uses in its cars, and Ford can reply: Quebec. What was the origin of the aluminum ore in Quebec? Ultimately, the aluminum or its components is a given. First catch your hare.

The second question, as Taylor notes, is morality. Machines are not moral or immoral. They are amoral. The guillotine is a machine for chopping off heads. I think it is a very bad thing. I suspect that most of the inhabitants of the state in which I live, Florida, think it a very good thing. Their only regret is that executions are no longer public. Science does not speak to morality. That is, it doesn't speak to moral foundations. It can enter moral decisions. Now that science has yielded the atomic bomb, we have the moral question of whether it should ever be used. The bomb as such does not answer this question. Harry Truman said yes. Philosopher Elizabeth Anscombe (1957) said no. You may think I am being a little bit shifty here, because I am well known for arguing that Darwinian evolutionary biology shows that morality at the level of directives—what philosophers call substantive or normative ethics—has no foundation—what philosophers call metaethics (Ruse 1986). This is true, but I think it is still open to ask why it is that nature is such that ethics is needed. More on this in a moment when we get to Christianity.

The third question is consciousness. I am right with Taylor and, before him, Gottfried Leibniz in the *Monadology*. Machines don't think.

> One is obliged to admit that *perception* and what depends upon it is *inexplicable on mechanical principles*, that is, by figures and motions. In imagining that there is a machine whose construction would enable it to think, to sense, and to have perception, one could conceive it enlarged while retaining the same proportions, so that one could enter into it, just like into a windmill. Supposing this, one should, when visiting within it, find only parts pushing one another, and never anything by which to explain a perception. Thus it is in the simple substance, and not in the composite or in the machine, that one must look for perception. (Leibniz 1714, 215)

Philosophers like Daniel Dennett (1992), who claim that once you have given a physical description of what is going on you have solved the problem, strike me as between silly and dishonest.

Taking computers a bit too seriously. I have more time for neuroscience-friendly philosophers like Paul Churchland (1995) and Pat Churchland (1986), who claim that more physical research will solve the problem. I doubt it, however. The origin-of-life question is unsolved, but I know what a solution would look like (Bada and Lazcana 2009). Crystallization or some such thing. In the case of consciousness, I don't even know what a solution would look like. In many respects, I follow philosopher Colin McGinn (2000) and am a "new mysterian," doubtful that we have the capacity to solve this (genuine) problem. More recently, as you will learn, I have been drawn increasingly to what is known as panpsychism, seeing mind in a monistic way as part of matter—or matter part of mind (Ruse 2017b). But even if this works better than some of the alternatives, like Cartesian dualism, I think this is a philosophical answer and not a scientific one.

Finally, there is the fourth question, that of ultimate meaning. You might think that this shouldn't be here. After all, the whole point of machines—of artifacts—is that they are made for something. So is it legitimate for—expected of—a scientist to ask what is it all for? Not so fast. Even with metaphors, especially with metaphors, you can restrict the domain. If I say my love is a red, red rose, it is hardly expected that you will ask, "Was she raised on organic Knock Out rose plant food?" It is certainly legitimate to ask this of real roses. From personal experience I can tell you that our roses bloom as never before. However, as we saw in chapter 1, the success of science depended more and more on excluding these purpose questions—final-cause questions—and concentrating on machines as things that work according to ongoing unbroken law. Forget the ends, especially the ultimate ends, and focus on the mechanism. Even with an organism, other than those designed by us, you don't ask what is the point? What is the point of a dandelion? A dandelion just is, and now spend your time trying to work out things like its methods of reproduction. I am not surprised that Weinberg was disappointed. As a scientist, he wasn't asking the question in the first place.

The machine metaphor shows that there are unanswered questions in science—genuine questions that science does not even attempt to answer—and by thinking on the nature of metaphors, we can see why.

Does Christianity Answer the Questions That Science Doesn't?

You can see where this argument is going. Turn now to Christianity. It can and does offer answers to all these questions. Why is there something rather than nothing? Because an all-powerful and all-good God freely created the universe and everything in it, including our planet and us humans. What is the foundation of morality? The will of God. What is consciousness, sentience? Being made in the image of God. What is the meaning of it all? Salvation, eternal life, and joy with our Creator.

Of course, there are some time-honored questions about these answers, but note that the questions must be philosophical or theological. They cannot be scientific. The Creator God must in some sense be a necessarily existing Being. Thomas Aquinas made it very clear that it cannot be part of the regular causal chain, or one simply asks: What caused God? (Ruse 2015). God is cause of Himself (or Itself). Although Dawkins (2006) dismisses it with a sneer, Anselm's ontological argument speaks to this. God is necessary, because His denial is contradictory. I suspect most theologians today go less with a mathematical or logical notion and more with the idea of God as self-generating—what is known as aseity. I am not sure how much sense it makes, but I don't think it a stupid idea.

The same with a Christian approach to morality. On being presented with the divine command theory of the foundations of morality, most trot out the Euthyphro problem. Is that which is good good because God wants it, or does God want it because it is good? Could God make it okay not to help little old ladies across the road and to mark up library books with yellow highlighters? Natural law theory speaks to this issue, defusing the problem (Quinn 1978; Ruse 1988b). God made us in certain ways, and so morality follows naturally from this. Children need parents, parents want children, so God made it our duty to love and cherish our children. Those dreadful parents who kept even their grown-up children all locked up and chained were immoral because they were unnatural. God wants us to be good, but what is being good is a function of what we are. God couldn't have made us and then made it acceptable to rape your own daughters. Morality can have a God-foundation.

I am not sure how helpful it is to be told that we are made in the image of God. It certainly doesn't tell us what consciousness is. But this latter is a scientific question. In the context of Christianity, being told that we are in the image of God tells us a lot. For a start, we have intellectual and moral obligations. We are not just dumb brutes. That is what made Adolf Hitler so evil. He was not an out-of-control elephant in musth. Most important, for whatever reason, God so cared for us as conscious beings—and not just blocks of wood or unthinking warthogs—He joined us here on earth and died on the Cross for our sakes. Hardly science, but (even if one finds them quite unacceptable) not obviously stupid answers.

Finally, meaning. I am going to have some questions about all of this later, particularly about eternal bliss with the Creator. The point I make here is that I don't think it illicit to speculate—if that is the word—about a heavenly future, whatever that might mean. And it does seem to me to give a meaning to life, not just then but now. We are not put on this earth just to have a good time but to make full use of our talents. God makes this demand and rewards us accordingly. "Well done, thou good and faithful servant: thou hast been faithful over a few things, I will make thee ruler over many things: enter thou into the joy of thy lord" (Matthew 25:21). Or otherwise. "Thou wicked and slothful servant, thou knewest that I reap where I sowed not, and gather where I have not strawed" (25:26). Of course, there are issues about what one ought to do with the talents, but that is another matter. We have a job to do, and it gives meaning. "We need to be sustained in a belief in the ultimate resilience of the good; we need to live in the light of hope. Such faith and hope, like the love that inspires both, is not established within the domain of scientifically determinate knowledge, but there is good reason to believe it is available to us through cultivating the disciplines of spirituality" (Cottingham 2003, 104).

Why Not Be a Christian? Some
Methodological Issues

So why not take this option? I have said that science cannot dis-
prove us here but that philosophy and theology can speak to the
plausibility of Christianity. It is at this point that I hit the barriers.
Before I start giving reasons why I turn from the religion of my
childhood, with both regret and relief, let me give first a word
about my stance or methodology. "Now faith is the substance of
things hoped for, the evidence of things not seen" (Hebrews 11:1).
I am totally with Søren Kierkegaard and, in the last century, Karl
Barth in thinking that religious belief is ultimately a matter of
faith. This is not a new thought: "Then saith he to Thomas, Reach
hither thy finger, and behold my hands; and reach hither thy hand,
and thrust it into my side: and be not faithless, but believing.
And Thomas answered and said unto him, My Lord and my God.
Jesus saith unto him, Thomas, because thou hast seen me, thou
hast believed: blessed are they that have not seen, and yet have
believed" (John 20:27–29).

All the great theologians have endorsed this at one level or another. Aquinas thought that reason can get there in the end: "For
certain things that are true about God wholly surpass the capability
of human reason, for instance that God is three and one: while
there are certain things to which even natural reason can attain,
for instance that God is, that God is one, and others like these"
(Aquinas 1259–1265, 5). Note, however, that reason—where
we could be wrong—is limited, and in the end, faith—where we
cannot be wrong—is top dog. "The truth of the intelligible things

of God is twofold, one to which the inquiry of reason can attain, the other which surpasses the whole range of human reason" (7). Ultimately, without faith you only get part of the story, and Aquinas makes clear that faith trumps all—how else could the ignorant or stupid or lazy get knowledge of God? John Paul II, in his encyclical *Fides et Ratio*, affirmed this position strongly: "The results of reasoning may in fact be true, but these results acquire their true meaning only if they are set within the larger horizon of faith: 'All man's steps are ordered by the Lord: how then can man understand his own ways?' [Proverbs 20:24]" (John Paul II 1998, 16). This ordering of faith and reason is not fortuitous. It is the result of original sin. "According to the Apostle, it was part of the original plan of the creation that reason should without difficulty reach beyond the sensory data to the origin of all things: the Creator. But because of the disobedience by which man and woman chose to set themselves in full and absolute autonomy in relation to the One who had created them, this ready access to God the Creator diminished" (22). Although someone like Plantinga is (let us say) somewhat Calvinistic about these matters, his is the absolutely central position.

What then of reason, what of natural theology and the like? For the believer, the right position is that of John Henry Newman: "I believe in design because I believe in God; not in a God because I see design" (1973, 97). As a Christian, one believes on faith all about the Christian God, and then one fleshes this out by looking at the world and using reason. After all, that is what being made in the image of God is all about. What one does not do is strive to get to God through reason— or bust. This is true even—especially—of someone like Aquinas.

To behave in this way is to reveal that you really don't think faith is good enough, and you rely on the fallible methods of reason rather than the certain truths of faith. John Paul II is unequivocal: "the truth made known to us by Revelation is neither the product nor the consummation of an argument devised by human reason. It appears instead as something gratuitous, which itself stirs thought and seeks acceptance as an expression of love" (1998, 15).

Today, this admonition applies particularly to those who try desperately to keep alive the Anglican tradition, ignoring the Calvinist theology of the Thirty-Nine Articles, filling the gap in the design argument brought on by the *Origin of Species*. Physicists rush in where biologists fear to tread. The so-called anthropic principle supposedly shows that the constants of the universe could not have been other than they are and have produced life (Barrow and Tipler 1986). They are "fine-tuned." Ergo God. Truly, it doesn't prove anything about God, but it does prove a lot about physicists (Ruse 2017b; Stenger 2011). Serious thinkers on the topic, like Weinberg (1999), are far from convinced that the fit need be that tight. In any case, it is ludicrous to argue from just one example, namely us—think of a number, double it, and the answer is a half. Not to mention multiverse hypotheses, which suggest that there will be an infinite number of possibilities and so our world was bound to come up sooner or later. More important, the whole exercise is theologically shabby and, in the light of Christian belief, nigh heretical. For all that in the years after the Reformation they may have helped the English define their own national identity, the arguments to God don't work that way and should not work that way.

How does the non-believer like myself handle all this faith business? I don't believe in the Christian God. I don't have faith. What then? Let me say straight out that I do not at once reject faith because it is not reason, thinking that it must be something weak or stupid or fattening. Take falling in love. David Copperfield, in Charles Dickens's novel of that name, probably still in his teens, is articled to become a proctor, a kind of lawyer who deals with a weird combination of issues, including wills and misbehaving clergymen and (of all things) nautical matters. One weekend, David is invited down to the home of the (widowed) senior partner, where he meets the partner's daughter.

> We went into the house, which was cheerfully lighted up, and into a hall where there were all sorts of hats, caps, great-coats, plaids, gloves, whips, and walking-sticks. "Where is Miss Dora?" said Mr. Spenlow to the servant. "Dora!" I thought. "What a beautiful name!"
>
> We turned into a room near at hand . . ., and I heard a voice say, "Mr. Copperfield, my daughter Dora, and my daughter Dora's confidential friend!" It was, no doubt, Mr. Spenlow's voice, but I didn't know it, and I didn't care whose it was. All was over in a moment. I had fulfilled my destiny. I was a captive and a slave. I loved Dora Spenlow to distraction!
>
> She was more than human to me. She was a Fairy, a Sylph, I don't know what she was—anything that no one ever saw, and everything that everybody ever wanted. I was swallowed up in an abyss of love in an instant. There was no pausing on the brink; no looking down, or looking back; I was gone,

headlong, before I had sense to say a word to her. (Dickens 1850, 390)

Not much reason here. Yet so human, so very human. And if this doesn't remind you of Hick on the top of the omnibus, I don't know what would. One thing that people of faith stress is that although it has a propositional content—God exists—there is much more. It "seeks acceptance as an expression of love" (John Paul II 1998, 15). The sense of being swept up by a force stronger than oneself—a force that is entirely good and beautiful.

I don't remember who was there, except Dora. I have not the least idea what we had for dinner, besides Dora. My impression is, that I dined off Dora, entirely, and sent away half-a-dozen plates untouched. I sat next to her. I talked to her. She had the most delightful little voice, the gayest little laugh, the pleasantest and most fascinating little ways, that ever led a lost youth into hopeless slavery. She was rather diminutive altogether. So much the more precious, I thought. (Dickens 1850, 391)

Is this enough? We've all had experiences of falling in love. Makes very good biological sense. Shakes up the genes. But what about faith where someone like me—perfectly normal, perfectly healthy, at David's age just about as daft and susceptible—just doesn't have it? Something that I am assured is perfectly open for me to have. It is not like I am suffering from a chemical imbalance that makes it impossible for me to fall in love. "Dora is the most beautiful woman in the world." "Dora is not the most beautiful woman in the world." "God exists." "God does not exist." We'll give David the first. Why shouldn't we give him the third? Because if we give him the third, he is going to insist that we reject the fourth. Whereas if we give him the first, he is probably very glad to give us the second. No competition. The point is that falling in love with Dora is a subjective thing. Falling in love with God is supposed to be something more.

Should I therefore conclude that I am like someone blind, who is missing one of the very real and genuine senses? I don't see blue because I am physically blind. I don't see God because I am spiritually blind. Or do I say that believers are deluded into self-deception because, as William James pointed out (1902), it is all very comforting? Here following the example of blindness, it surely is legitimate to turn to reason and evidence, natural theology in some form. Why does the physically blind person take seriously the claim that he or she is lacking a sense and the sky really is blue? Well, you can feel objects, for a start. You can smell them. You can hear them smash when they fall. Some taste nice or nasty. Another sense, especially one that people can show you that they use, is more than plausible. If it isn't, why do you have those holes in your face? This all said, claim for yet another sense—a sixth sense—is a lot more iffy. What's it like? How does it function? How do humans use it? What part of the body does it emanate from? No good answers. No strong reason to believe in it. Really suspicious if people have a strong psychological reason to believe in it.

Same with spiritual blindness. I don't have a sense of absolute helplessness caught up in an overwhelming force of love and affection. I am not saying that faith must be sparked by *un coup de tonnerre*, as it was for Paul and John Hick and David Copperfield. Many write of struggling and resisting and only slowly being overcome.

> You must picture me alone in that room in Magdalen, night after night, feeling, whenever my mind lifted even for a second from my work, the steady, unrelenting approach of Him whom I so earnestly desired not to meet. That which I greatly feared had at last come upon me. In the Trinity Term of 1929 I gave in, and admitted that God was God, and knelt and prayed: perhaps, that night, the most dejected and reluctant convert in all England. (Lewis 1955, 115)

Why should I think you have it and that what you believe is absolutely true? If there are reasons, natural theology pro or con, that bear on this, then it is proper—mandatory—to consider them. If we walk into a room and you can tell me at once that there is a certain object in the middle, which I can only feel as I grope around, you have reason (and empirical evidence) on your side. I am right with John Locke on the need for Christianity to be reasonable. This is not the Archdeacon Paley position, where you use reason and evidence to justify your faith but, when you don't have faith, using reason and evidence to ask if you are missing something. Remember, you are not (yet) committed to the idea that reason has been corrupted by original sin, so it is still legitimate (mandatory) to use reason. Here are three critical reasons I take very seriously.

Why Not Be a Christian? Evil

The first reason is the venerable problem of evil. If God is Creator and all-good and all-powerful, how come there is evil in the world? Usually the question is divided into two. How come there is moral evil in the world? Heinrich Himmler. How come there is natural evil in the world? The Lisbon earthquake. These days, my impression is that Christian philosophers tend to be rather smug. They feel they have got the problem licked. Free will speaks to moral evil, and necessity speaks to natural evil. To be made in the image of God—the best of all possible choices—we had to be free and hence the possibility of moral evil. To create, God had to make decisions—the best of all possible worlds—and that is why burning hurts, because it brings on instant action away from a significant danger.

Paradoxically, Darwinian evolutionary theory goes some considerable way to support these arguments (Ruse 2001a). With respect to the free-will problem, having some form of flexibility is a very sophisticated and powerful adaptation that we humans have. We can live in varied and complex situations, adjusting rapidly when things change. If we were programmed like the ants, we could not do so. It rains, and the pheromone trails get washed away, a lot of foraging ants get lost and die. Mother ant compensates by having thousands of children, so if some go, there are plenty more. Humans are not like this. Imagine if someone asked you how many kids you had. "I have three. No, one just went to McDonald's, and it's been raining. Better say two." That's a joke, because we are built along the lines of a Mars rover. It can deal with obstacles without outside instruction. We are the same. As opposed to ants, which have gone the route of so-called r-selection, producing many offspring cheaply and letting nature take care of itself (and us), we have gone the route of so-called K-selection, having few offspring because they need more energy to produce and taking care of nature ourselves (MacArthur and Wilson 1967). We need to be flexible, in the sense of being able to reassess in the light of circumstances, and the reason for this is that—undoubtedly, part cause and part effect—we are very social, and being social means being flexible. If opportunities come up, you have to make decisions. Am I going to get involved in hunting and killing this giraffe? If I do so, what am I going to expect in the way of a share of the meat? And so forth. No one fixed answer but to be worked out according to circumstances.

As far as natural evil is concerned, Dawkins (1983), of all people, has argued that the only way you can, naturally, get design features is through natural selection. Lamarckism is false, and other alternatives, like saltations (jumps from one form to another), simply don't lead to adaptations. They exemplify Murphy's Law. If it can go wrong, it will go wrong. Unless there is a powerful counter-force, like natural selection, organisms won't work. Saltations lead to things like dwarfism or idiocy—in one leap or generation. So this means that at least some natural evil is indeed part of the Leibnizian trade-off. You want humans? Then you had better put up with fish eating fish, with birds eating rabbits, and with viruses causing all sorts of horrendous diseases. Natural selection doesn't explain all natural evil—the Lisbon earthquake—but it takes a big bite out of it, to use a pertinent metaphor!

So far, so good. But not good enough. I am with Fyodor Dostoevsky on this one—or at least with Alyosha. Ivan asks his brother a question.

> "Tell me yourself, I challenge your answer. Imagine that you are creating a fabric of human destiny with the object of making men happy in the end, giving them peace and rest at last, but that it was essential and inevitable to torture to death only one tiny creature—that baby beating its breast with its fist, for instance—and to found that edifice on its unavenged tears, would you consent to be the architect on those conditions? Tell me, and tell the truth."
>
> "No, I wouldn't consent," said Alyosha softly. (Dostoevsky [1879–80])

The free will of Hitler and Himmler and the rest of that sorry crew does not outweigh the deaths of Anne Frank, Sophie Scholl, and Dietrich Bonhoeffer. Nor does it help to look down the road and say God will make it all right in the end. There may be a God. Not the Christian God. Thank goodness. I don't want the Christian God to exist.

Why Not Be a Christian? The Problem of Pluralism

Second, there is the problem of different faiths. I was born and brought up as a Christian. Had I lived in a Polish ghetto before the war, I would be a believing Jew. Born in Saudi Arabia, a Muslim. In Tibet, a Buddhist. In India, a Hindu. In California, a pagan. And so it goes. Why is my religion fortuitously right? I see two strategies here for saving Christianity, having more than (let us say) a one-in-a-hundred chance of being right. Someone has to win the lottery to get you into heaven, but why me? The first strategy, "religious exclusivism," simply says that my position is right, or at least there is no presumption that my position is worse than others, so go with the devil you know rather than the devil you don't know. Plantinga (2000a) takes a line like this—although note that I take this to be the position of most, if not all, Christians. Remember that direct insight into God, a *sensus divinitatis*, is in some sense is self-validating. Plantinga thinks that other believers are corrupted by original sin and think falsely. Why would he change his thinking? He knows no argument that could do this, nor does he expect one. He is in much the same position as G. E. Moore (1925), who, challenged about the existence of the external world, held up his hand and said, "I refute him thus" (or some such words). Any alternative to the hand existing, to God not being the Christian God, simply is not plausible.

I suspect that if pushed, many people, including Christians, find this attitude grotesquely ethnocentric. From the same stable as white-supremacy thinking. Which leads to somewhat of a paradox, because this kind of thinking is central to Christian belief and practice, if not always put in the kind of stark way of Plantinga. If you are going to soften your stance, what's the alternative? "Religious pluralism," where you argue that all religions are in some sense true, all trying to get an insight into ultimate reality. That there are different ways is very much what you would expect from human cultures. Different languages all essentially doing the same thing, communicating—"I love you," "Je t'aime," "Ich liebe dich"—and different religions equally essentially doing the same thing, explaining—the world is more than materialism, humans are special, we have obligations. It is like painters trying to capture a landscape. John Constable and Vincent van Gogh would give us very different versions of the same place, but in the end, it is the same place, and that is the cause of it all. You can push the analogy a bit more. I suspect most people would say that van Gogh is better than Constable, truer in some sense. I, however, English-born, adore Constable, and whenever I'm in London, I try to get an hour in the National Gallery to look at *The Hay Wain*. It all comes down to culture.

Religious exclusivism and religious pluralism are usually presented as non-overlapping alternatives. My suspicion is that many Christians are exclusivists in name and pluralists in feeling and action. Hard-line exclusivism is morally offensive, so some infusion of pluralism is morally attractive. Judaism has things out of focus in respects, but to say that Jews are not in a direct relationship with the God of Christianity is simply offensive and wrong. Just so. Unfortunately, for all that it is warm and cuddly, religious pluralism—pushed strongly by Hick (1980), who had an enviable track record of working with interfaith groups in the city of my birth, Birmingham in the British Midlands—still has the epistemological problems. Most significantly that it is hard to see what is left in common once you start to strip away the cultural accretions. Ultimately, there is no getting away from the fact that Jesus, as the son of God—as God Himself—goes quickly once you turn to Judaism and Islam, religions that supposedly worship the same

God as Christians. Indeed, God Himself goes once you turn to the Unitarians, let alone the Buddhists. And then, going the other way, what price reincarnation? What remains? Hick writes:

> Let us begin with the recognition, which is made in all the main religious traditions, that the ultimate divine reality is infinite and as such transcends the grasp of the human mind. God, to use our Christian term, is infinite. He is not a thing, a part of the universe, existing alongside other things; nor is he a being falling under a certain kind. And therefore, he cannot be defined or encompassed by human thought. We cannot draw boundaries around his nature and say he is this and no more. If we could fully define God, describing his inner being and his outer limits, this would not be God. The God whom our minds can penetrate and whom our thoughts can circumnavigate is merely a finite and partial image of God. (1973, 139)

We are all like the blind men and the elephant. We describe different parts of the elephant as we feel it, but in the end, it is all one elephant, and our knowledge is limited.

A nice picture, but will it do? Even if you give Hick all of this—and, as I have said, he is not as much out of the Christian tradition as some of his critics suggest—we still have something there. Hick thinks it is good. But what about the pagans? Hick says he is not talking about materialism or humanism. The pagans strike me as far from materialism and humanism as it is possible to be. They are deeply spiritual, thinking the earth in some sense an organism and our obligations starting there. Listen to Peaseblossom in my dialogue (Ruse 2016). But I am not sure they always think the ultimate forces are necessarily good. These forces are in some sense indifferent, and we must work accordingly. Like the God of Job, it is not for us to question. It is for us to obey. However, even if you ignore the pagans, things are stretched. Was Jesus the son of God or not? It is no trivial matter to the Christian. I don't think the difference here between, say, Christians and Jews is just a matter of styles—Constable versus van Gogh—reducible to different cultures. So, in the end, I am not convinced, and not having Plantinga's sense of certainty, I regretfully pull back from all religions. A bad case of original sin, I guess.

Why Not Be a Christian? A Marriage in Trouble

Finally, for me, the killer (Ruse 2015). All agree that Christianity is a fusion of Greek thought and Jewish thought. On the one hand, the influence of Plato through Plotinus on Augustine and the influence of Aristotle on Aquinas. Something central to Catholicism. On the other hand, the Bible, the Old Testament and the New Testament. Something central to Protestantism. The trouble is that the fusion doesn't work. The Greek conception of God is very much like the Platonic Form of the Good, something outside time and space from which all else flows. Augustine in his *Confessions* is clear on this:

Thy years neither go nor come; but ours both go and come in order that all separate moments may come to pass. All thy years stand together as one, since they are abiding. Nor do thy years past exclude the years to come because thy years do not pass away. All these years of ours shall be with thee, when all of them shall have ceased to be. Thy years are but a day, and thy day is not recurrent, but always today. Thy "today" yields not to tomorrow and does not follow yesterday. Thy "today" is eternity. (Augustine 397–400, 396)

Same for space.

The Jewish conception of God is God as a person. From the start: "And they heard the voice of the Lord God walking in the garden in the cool of the day: and Adam and his wife hid themselves from the presence of the Lord God amongst the trees of the garden" (Genesis 3:8). And this continues right through into the New Testament. Take the parable of the Prodigal Son. We have here a father with two sons, who loves them both. The younger son clears off, finally comes home, penniless, and is welcomed by the father. The older son, understandably, is a bit cheesed off. He has been stable and faithful all along, but the fatted calf is not being killed for him. The father understands this and counsels and consoles him wisely. And if this isn't a person, I don't know what is. If you say it is all metaphorical, unpack the metaphor, and get away from God as a person. Which, of course, you cannot do and which Protestants don't want you to do. Richard Swinburne makes it clear that this is the very beginning of their faith: "Theism postulates God as a person with intentions, beliefs, and basic powers" (2004, 7). You cannot say it more fairly or more explicitly than that.

Beings outside time and space cannot have the emotions and feelings of persons. These are things necessarily in time and space.

The younger brother is thoughtless and then sorry. The older brother is jealous. The father is caring, happy, understanding. Even if you say that God's necessity is not the necessity of a right-angled triangle, that is about as close as we can get. Plato in the *Republic*, remember, had the objects of mathematics above the line, snuggling up next to the Forms. And right-angled triangles just don't have emotions. That's the payoff. No going bad. No going anywhere. A point that some of the savvier Christian theologians saw fully. Whatever is the personhood of God, it simply cannot be like ours. Thus Anselm: "For when thou beholdest us in our wretchedness, we experience the effect of compassion, but thou dost not experience the feeling" (1077–1078, 13). Thus Aquinas: "To sorrow, therefore, over the misery of others does not belong to God" (1265–1274, I, 21, 3).

I want no part of a God who feels no compassion for Anne Frank or no sorrow at the sufferings of Sophie Scholl and Dietrich Bonhoeffer. No thanks. A conclusion I know will leave my many good Christian friends simply shaking their heads—sorrowfully perhaps, compassionately certainly—but completely unmoved. The factors I take as pointing strongly to my realistic take on life they take the other way. Ultimately, though, it is a matter of commitment. My Christian friends have faith in their beliefs. "For I know *that* my redeemer liveth, and *that* he shall stand at the latter *day* upon the earth" (Job 19:25). That, for them, is definitive. Remember Hick on his moment of conversion. Not all have had this experience, but many have. Keith Ward, sometime Regius Professor of Divinity at the University of Oxford, who takes joy in hauling me over the coals publicly, had an experience of the Christ as a young man, and that was it.

I am a born-again Christian. I can give a precise day when Christ came to me and began to transform my life with his power and his love. He did not make me a saint. But he did make me a forgiven sinner, liberated and renewed, touched by divine power and given the immense gift of an intimate sense of the personal presence of God. I have no difficulty in saying that I wholeheartedly accept Jesus as my personal Lord and Saviour. (Ward 2004, 1)

I just haven't had anything like this. I just haven't. I tried when I was younger, but it just wasn't there. And that, I suppose, is where we part company. I don't have faith, and reason doesn't lead me to believe that I am missing out on something. Christianity does not and cannot give me the ultimate answers.

The Buddhist Alternative

In his relentlessly detailed discussion of the meaning-of-life question, South African philosopher Thaddeus Metz distinguishes between God-centered theories of meaning and soul-centered theories. We have been dealing in Christianity with a God-centered theory of meaning. Without God, life is meaningless. "Soul-centered theory is the view that significant existence is nothing but being constituted by a soul that lives forever in a certain way, where a soul is an indestructible, spiritual existence" (Metz 2013, 123). Naturally, in this context one thinks of Buddhism, although there might be questions (to be raised) about souls, not to mention life "forever." I certainly don't pretend to know this religion as well as Christianity, but I lived through the 1960s, so Buddhism is not entirely an unknown entity! To contrast with Christianity, let us see where Buddhism leads us (Harvey 1990; Edelglass and Garfield 2009; Skilton 2013).

Buddha is a description meaning "Enlightened One" and strictly speaking is not a name. There have, in fact, been many buddhas, and there will be more to come—at least, we should hope so fervently, because that is the path toward which we all tread. The Buddha, the founder of the religion that has the name, was a prince, Siddhārtha Gautama, who lived in the northeast of India around four to five hundred years before the birth of Christ. (There is dispute among scholars as to the exact dates.) Paradoxically, the religion today is not in India much but in northern countries like Tibet, China, and Japan and right down below in Sri Lanka (Ceylon). There are one or two interesting parallels with Jesus, starting with a miraculous birth. Like Jesus, the Buddha left home (he was about twenty-seven, a year or two younger than Jesus)—despite his father's best shielding intentions, he had become aware of disease, old age, and death—and after a period of training (including a later-rejected time of fasting), he started preaching, attracting many followers and disciples. Also, as we shall see, much of the teaching stressed the importance of leading a moral life, and it was put in an eschatological framework of improvement for a later state of being. There is also a temptation incident, involving the evil god Māra.

The differences with Jesus, however, are striking, starting in the Buddha's case before his birth, with white elephants entering his mother's body. Somewhat more prosaically, the Buddha had a wife and a son, whom he left when he felt the call. The Buddha was welcomed by authorities and lived a long life, dying peacefully at the age of eighty, surrounded by his followers. Most important, though, is the fact that the Buddha, although very special, made no claims to divinity, most especially not to a Creator God. Perhaps best known of all about Buddhism is that it has no such God. The world just is, always was, always will be. Or rather perhaps we should say the worlds just are, for it is believed that there are many universes such as ours and that systems like this on earth—animals, plants, humans—are constantly being created and then fading away. There are gods, but these are not creator gods. They are part of the natural hierarchy. Like Christians—Ten Commandments, Trinity (three in one), seven deadly sins—Buddhists are into lists. There are five or six levels of being. At the bottom is hell—or perhaps better, Purgatory, for it is not a permanent home. Above this are petas, wraiths, rather like the Wilis, or jilted maidens, in the second act of *Giselle*. Animals above this, not plants apparently, and then humans. Then there are gods, divine beings, including lesser gods, asuras who are not that nice, and greater gods, who are nice, and the devas, including the greatest gods, the brahmas. One, the Supreme Brahma, has a mistaken idea about his creative status in the world. Interestingly, the Buddha knows better, illustrating that although humans are below the devas (they are above the asuras), in respects they are or can be superior.

Buddhist ethics centers on the belief in rebirth—reincarnation—for it is an essential part of the teaching that all is transitory and that this life is but a passing one among literally millions. Finish one life and start another. (There is some debate about whether it is immediate or whether there can be a gap in

time.) All that is essential is how you do or behave in this life, for
it has implications for the next life, the level at which you will be
born—from human to animal, say—although it is made clear that
the baggage you bring and its effects, karma, are not necessarily
immediate. Hitler, for instance, might be working things out now
at the codfish level, but even if he does well—sacrifices himself to
the trawler's net for the sake of the shoal—he most probably will
have many more lives of making up to do. Hence, although it is
an important part of Buddhism that we are free—we get credit
or not for our acts—we are also in a sense bound or fated by our
past. We get a remarkable echo of this in Thomas Hardy's *Tess of
the D'Urbervilles*. Tess kills her lover freely. She tells her husband
that she did it because she knows he cannot love her fully while
the man still lives. And yet Hardy makes it clear that some of Tess's
aristocratic ancestors were truly vile and that Tess, both in looks
and in temperament, carries the taint. A kind of Augustinian original sin (Ruse 2017a).

The overall aim is to move upward. As I noted, not necessarily
to become a god but, like the Buddha, to have reached the ultimate
goal, at which point one moves off the continuous escalator (one
that goes down as well as up), and achieves the state of enlight-
enment, nirvana (or nibbana). It seems, therefore, even before we
get to the nature of right action, we are dealing with a soul-based
religious explanation of meaning. This is true but must be quali-
fied. The Buddha stressed that there is no such thing as the self, as
such, a kind of thing—Cartesian *res cogitans*. In the words of the
Buddha, "all phenomena lack any underlying and permanent substance; they all have the characteristic of 'no-self'" (Mitchell and
Jacoby 2014, 35). There is, of course, something there. The Buddha

could speak happily of "myself" or "yourself," but truly one should think of a collection of parts, always in change. These five parts are first the material body, second sensation or feeling, third sensory and mental objects (ideas or concepts), fourth the motives leading to action (like joy and hatred), and fifth and finally consciousness. These make for the "not-self," anatta (in Sanskrit anātman). Not-self in the sense of always fluid but fairly considered a sense of self if one thinks of continuity—especially what is passed from one life to the next. Hitler isn't going to get off that easily.

What then is the moral life that points to nirvana? The key is that this is a life of suffering, of discontent—dukkha. This is not just physical suffering—sickness, aging, death—but also mental and a general lack of total satisfaction. Buddhists don't deny happiness, but it is always transitory, not lasting. Dukkha therefore impinges on us at all levels of the not-self—unpleasant experiences, thoughts, and the rest. Getting yourself stuck in an awful job or family, for instance. This leads straight into the Four Noble (or Holy) Truths, given by the Buddha to his disciples in his first sermon. Based on the practices of the physicians of his day, he started with the problem—namely that this is a life of suffering, of dukkha. Then second, the reason for the problem—ultimately it is all to do with desire, with craving. This gives rise to the three great faults of greed, of hatred, and of ignorance or delusion. Note the contrast with Christianity, where it was disobedience and pride that were the problem. One never gets the sense that Adam and Eve cared too much about the nature of the fruit they were gobbling down. Third, the solution to the problem: achieve nirvana, where you are beyond dukkha. Fourth, of importance to us—the Buddhist equivalent of the Beatitudes—how to achieve the solution, through following the eight ways (the Eightfold Path) to righteousness.

Two questions, obviously. First, what is nirvana? It is not nothing—a kind of eternity of non-existence, dreamless sleep, as Plato offered as one option in the *Phaedo*, the story of Socrates's last day on earth. But it is hard to say what it is. We are told that unlike existence now, it is "unborn, not-become, unproduced, not-compounded." What can we say positively? Part of the very point is that we can't, but sometimes metaphors are used. It has been called "coolness" (as in a cave), a powerful and attractive thought given the heat of the summer in India (Mitchell and Jacoby 2014, 46). One presumes it is a kind of higher transformed consciousness, and that is how it has been presented. "The consciousness in which nothing can be made manifest (like space), endless, accessible from all sides (or: wholly radiant)" (Harvey 1990, 63). Second, how to achieve it? The first two of the eight ways are to do with wisdom—getting the right ideas and getting the right thoughts, overcoming ignorance. The next three are to do with moral virtues—right speech, right action, and right livelihood, living in a proper way. The final three are to do with cultivation of the mind, where, as is well known, meditation plays a big role. These three are right effort, right mindfulness, and right concentration.

As is also well known, Buddhism puts a great value on the sanctity of life—of all animal life—and one finds pacifism and vegetarianism of various degrees of strictness. Expectedly, also as Buddhism grew and developed, there were different flavors of right action. Earlier Buddhism, Theravada Buddhism—today "Southern Buddhism"—found in countries like Thailand and Sri Lanka, put much emphasis on the improvement of self. Later

Buddhism, Mahayana, from about the time of Christ—in places like Tibet, China, and Japan (the variant Zen, found in China and Japan, came about a thousand years later)—puts great emphasis on reaching out to others. One is reminded of the teaching of Paul, which stresses personal salvation, justification by faith, as opposed to the teaching of Jesus (and Peter), which stresses help to others: "Verily I say unto you, Inasmuch as ye have done it unto one of the least of these my brethren, ye have done it unto me" (Matthew 25:40). The parallels are not too tight, and obviously no one is saying only self or never self. Or only help or never help.

Do We Get Meaning?

One could go on using up one of those apparently available eons describing Buddhism in ever greater detail. Enough now is on the table for our purposes. The first thing clearly to be said is that, buying into the story or not, assuming that you are prepared to say something positive about religion, one can say something positive about Buddhism. It merits thoughtful response. It is a sophisticated world picture, both with respect to cosmology and with respect to proper moral action. There are some significant differences with Christianity. However, we surely sense now that it would be silly to say simply that Buddhism is atheistic and leave it at that. Anything further from the world of the *God Delusion* would be hard to imagine. That said, although there are gods, there is no Creator God. It is perhaps for this reason, not fearing His wrath—thou shalt have no other gods but me—we find that people often follow other local religions as well as Buddhism. In Japan, people get married with Shinto rights and buried with Buddhist rights. As for Dawkins, however, the world just is—always was, is now, always will be. Without rhyme or reason. It just is.

You may find this unsatisfactory, but it gets you out of a lot of problems, starting with that of evil. Suffering happens. Get used to it. Get over it. Don't whine. Don't go groveling for help. It's your karma, no one else's. It also gets you out of problems that for me bedevil the Christian God. You are not trying to blend together the concepts of two very different philosophies and cultures. Is God eternal or a person necessarily in time? Doesn't matter, because there is no such God. Note, however, that with respect to religious pluralism, Buddhism is like Christianity in having at its core a sense of being that transcends human understanding and description. No doubt someone like Hick would say that is his starting point. Different religions do at their cores speak of the ineffable—truly good if not truly understood. Perhaps so, although note that here we have precisely the difference being highlighted between Christianity and Buddhism. Christianity is God-centered, and Buddhism is soul-centered, whatever that might mean precisely. Their ineffables seem not to be of the same thing, although I suppose one could say that is precisely what happens with ineffables. Who is to say that they are not the same in some fundamental sense?

So here then we have another approach to meaning, or Meaning. For Christianity, our lives have meaning because of the possibility of eternal bliss with our Creator. For Buddhists, our lives have meaning because of the possibility of nirvana, eternal bliss off the escalator of life and suffering. The sixty-four-thousand-dollar question is whether there is any reason to believe it. You might say that, true or not, Buddhism has much to commend it over Christianity, and I do confess to considerable sympathy with that judgment. I am not a vegetarian, as much because of cultural reasons as any. Given attachment to the heavy-handed postwar diet on which I was raised, I would find it difficult to relinquish toad in the hole, not to mention fish and chips. Thoughts of spotted dick—best made with suet—have me swooning. I guess I can have the mushy peas, although not pease pudding! I can respect vegetarians. Vegans push me a bit. No Roquefort? I think a slab of that is worth the risk of an incarnation or two as a Wili. The question remains: Is it true?

This is not a new question for me. In my teens, my mother having died, my father married a woman from Germany, whose family was totally committed—as my father soon became—to the ideas of the Austrian clairvoyant Rudolf Steiner, founder of the Waldorf schools. Influenced by Eastern religions, Steiner (1914), too, was deeply into the idea of reincarnation. At the age of sixteen, hormones pumping vigorously, I was solemnly assured that I had once been a woman and would again someday be a woman. I am not sure if there was direct cause and effect, but I stand by the decision that I came to then. It's a nice idea, but it just ain't true—and on reflection, it isn't such a nice idea. The empirical evidence is, to say the least, somewhat shifty. Those people who claim memories of past incarnations tend to be a bit morally smelly, and the same goes for those who triumphantly get small children to remember that they used to be the queen of Sheba. As Plato showed only too well in the *Meno*, a skilled interlocuter like Socrates can get a kid, the slave boy, to say practically anything he wants. I really don't quite know what it would mean to say that I used to be Queen Victoria. I can appreciate her hearty sex life, but Prince Albert? No, thanks. I share her fondness for dogs, but frankly, I can think of few things I want to do less than spend the holidays in the gloomy north of Scotland, surrounded by hairy types in skirts, playing instruments that should have been quietly throttled at birth. And as for being empress of India? Even translated into a male form, I don't see how once it could have been part of me. Evening after evening with bores at the club, downing endless gin and tonics, moaning about the quality of the servants. Not to mention amateur performances of Gilbert and Sullivan, with the regimental colonel playing Pooh-Bah—even worse, in drag, along with the vicar and the schoolmaster, playing a little maid from school.

You could say that I was born with a certain karma. Overall, I think I am a pretty decent chap—good and loving with my wife and hugely overindulgent toward my children and graduate students. (Vigorous nods of agreement from Dan, Liz, and Jeff.) I really am someone who works hard and takes seriously his responsibilities as a teacher. (Remember, I am the guy who taught you how to BBQ chicken wings so you would have some job skills in the Deep South when all else fails.) I am not sure I would have been very good had I been forced into being a soldier. There were one or two embarrassing moments in my childhood when I exhibited naked fear and screamed like mad. I remember an episode with a beehive and later an adult saying, "Well, we are all cowards sometimes." I certainly was, to my great shame. I just find it hard to relate this all back to a young soldier who died on the first day of the Somme or perhaps more plausibly to a young soldier who was shot at dawn for cowardice. We were both cowards, but was I really a coward because he was? Am I a dedicated teacher because Dr. Arnold of Rugby was? Do I love France because my karma was passed on to me by Madame de Pompadour? I went to Versailles once but never did I feel the need to relieve myself in the nearest free corner.

Don't misunderstand. I don't think rebirth is silly. I just don't think it is empirically true, and that is before you get into the kinds of continuity problems that philosophers so love. What would it mean to say I am the incarnation of Queen Victoria? I know I am the same Michael Ruse as the one who went to sleep last night, because I remember being Michael Ruse and I have Michael Ruse's body when I wake up. If, like in Kafka, I woke up as a giant insect, even I would have doubts. There are questions about whether I would be the same Michael Ruse if I were reincarnated on the Day of Judgment. Am I really the same person with a ten-million-year gap? I have seen it suggested that we could be like a CD or a program stored in God's mainframe. But then what is to stop God from making two Michael Ruses or ten thousand, for that matter? Dizzying thought. What is to stop my karma from being passed on to two people in the future, especially if I am a not-self? Perhaps in a year or two, as I peer up at you from the watery depths, I will find that I made a Big Mistake. For now, I doubt it—meaning, I really doubt it.

Religion offers meaning—Meaning. I don't think this is a stupid approach. I respect (and love) many who take it. It is not for me. Not because I am a Darwinian but because I just don't think it is true. A paragraph or two back, I joked about original sin. Are you going to say that it all comes down to the fact that I let the taint of the sin of Adam conquer me in a way that it has not conquered John Paul II and Plantinga and Hick and Ward? God in His Calvinist sovereignty picked me out for non-belief and hence for eternal doom and disaster? I joke that having had one headmaster in this world, I was damned (in more ways than one) if I was going to have another headmaster in the next world. That's not really the reason. My non-belief is mine and mine alone. Unlike these other men, I simply don't believe in the existence of God. I don't have faith. Am I, and I alone, doomed? A theological perspective not restricted to Calvinism: "because of the disobedience by which man and woman chose to set themselves in full and absolute autonomy in relation to the One who had created them, this ready access to God the Creator diminished" (John Paul II 1998, 22). I can but say: Don't be so condescending—and dangerous. That kind of thinking led to the Inquisition. Darwinism opened up the void. Religion doesn't fill it.

3 | DARWINISM AS RELIGION

"Naked came I out of my mother's womb, and naked shall I return thither: the Lord gave, and the Lord hath taken away; blessed be the name of the Lord" (Job 1:21). Well, natural selection has done its fair share of taking away. Can it also do some giving? Less metaphorically—although my quoting of the Bible was deliberate and, as you will see, contains a hook—if we do not take the religious route, and if natural selection seems to point to the bleak world of Thomas Hardy, is there any case for saying that natural selection can nevertheless contribute positively to an alternate world picture? A world picture that gives meaning to life? We shall see that there is. To tackle the problem properly, however, we need to make and follow a division made generally by those asking about the meaning of life. Religious approaches—Christianity and Buddhism are our examples—posit an extra-human reality. Followers are not extreme idealists, as George Berkeley would have been without his God—no small qualification, although of course, Berkeley's God was all spirit and not matter. What makes for meaning is not all in the mind. This is true even for a religious pluralist like John Hick. It is true that he is committed to allowing a relativity (cultural bias) about the divinity of Jesus, but he strongly affirms an underlying, good, ultimate life force, one that we may not be able to comprehend but is absolutely and completely real.

So let's say simply that religions take an objectivist approach to meaning. To advance on from this world, it is not your choice that you ought to do X, Y, and Z. It is an objective truth, because that is the way things are. So let us frame our first question in this mode. Can a natural-selection-governed or -inspired world picture give an objectivist understanding of meaning? Can natural selection impose upon us a set of rules for right conduct and point to a worthwhile end to which we should aspire and labor? Thinking of the contrary, our second question must be that if achieving an objectivist understanding proves impossible, and hence in some very real sense we are thrown back on ourselves, can natural selection contribute to such a subjectivist world picture? Can it help us to find meaning after all?

Can Darwinism be a Religion?

I will take these questions in turn, starting in this chapter with the objectivist approach. Let me now double back to my hook. Religion yields the paradigm of the objectivist approach. Can we put ourselves on the path to an answer by asking if, in some sense, a Darwin-inspired world picture can be or can function as a religion? In our opening lines, religion sets the pace for questions about natural selection. Is it worth continuing in this mode? Cutting at once to the quick, I argue that this is indeed the right way to go and gives the right questions to ask. From the time of the *Origin*, what I shall now by stipulation call Darwinism—restricting the term to a secular world picture inspired by Darwin's theory of evolution through natural selection—has existed and flourished. This holds down to—very much down to—the present (Ruse 2005; 2017a; 2018b).

At once I make a caveat. I am now distinguishing *Darwinism*, as I use it, from Charles Darwin's theory of evolution through natural selection, as it was in the *Origin* and as it has matured into today's theory. I take without argument that this latter was and always was, and is now, a genuine scientific theory, with a proper causal understanding of the physical world (Ruse 2006). Increasingly, it has been backed by empirical evidence, both from nature and from the laboratory. I know there have been various scientific challenges—genuine and pseudo—to the supreme place of Darwin's theory. I assume, however, that these are either inadequate or, more often, readily incorporated into the main theory or paradigm. More important here is that in the light of the arguments in chapter 2, because Darwinian theory is genuine science, no more no less, on its own it cannot answer our questions about m/Meaning. If such questions are to be answered in the Darwinian world, then something must be added that makes this possible. If Darwinian thinking is to be turned from straight science into a kind of religion, ask about the new ingredient.

Don't put the cart before the horse. First, ask whether there might be such a thing as Darwinism, a secular religion or substitute for such a religion. Many non-believers today, New Atheists like Richard Dawkins, deny with indignation the charge that they are in the religion business. To be honest, it is hard to take too seriously the protestations of someone who begins a book with: "The God of the Old Testament is arguably the most unpleasant character in all fiction: jealous and proud of it; a petty, unjust, unforgiving control-freak; a vindictive, bloodthirsty ethnic cleanser; a misogynistic, homophobic, racist, infanticidal, genocidal, filicidal, pestilential, megalomaniacal, sadomasochistic, capriciously malevolent bully" (Dawkins 2006, 1). If those aren't the words of an Old Testament prophet, I don't know what are. We need not, however, quibble about terms. Evolutionists themselves openly tell us that they are in the religion business.

Start with Thomas Henry Huxley, Darwin's great supporter—his "bulldog." Huxley, an almost lifelong university teacher and researcher, was keen above all to professionalize science, including his own field of the life sciences (Ruse 1996). As one who went very successfully into university administration—he was the dean of the new science-oriented college in South Kensington—he knew that for success, he had to attract students, and these had to have the promise of good jobs at the end of their courses. He sold anatomy and physiology to the medical profession, arguing that he could give the students the basic science, and then they, the medics, could do the clinical. An offer that was very gratefully received by a group that was itself trying to professionalize and improve, particularly after the absolute disaster of the Crimean War (Ashton 2017). He sold morphology to schoolteachers, arguing that in today's world, hands-on experience in science was far better training than the classics. To this end, he sat on the first elected London School Board and trained many teachers himself, including the future novelist H. G. Wells.

Huxley very carefully separated off this professional science from the stuff he preached night after night, in workingmen's clubs, on society lecture podia, and very much in hugely popular essays that he produced non-stop. A student, Father Hahn (a Jesuit), who studied with him in 1876, wrote: "One day when I was talking to him, our conversation turned upon evolution. 'There is one thing about you I cannot understand,' I said, 'and should like a word in explanation. For several months now I have been attending your course, and I have never heard you mention evolution, while in your public lectures everywhere you openly proclaim yourself an evolutionist'" (Huxley 1900, 428). Huxley knew full well what he was about. Evolutionary thinking simply does not cure a pain in the belly. Evolution, Darwinian evolution, has another function: to offer an alternative to the conventional Christian religion.

Huxley was quite explicit that he was seeking a new religion to supplant the old, Christian religion. Even before the *Origin*—he had been primed about Darwin's theory—he wrote about seeing (conventional) religion and science forever at war. "Few see it but I believe we are on the eve of a new Reformation and if I have a wish to live thirty years, it is that I may see the foot of Science on the necks of her Enemies. . . . But the new religion will not be a worship of the intellect alone" (quoted in Desmond 1997, 253). This was from a private letter, but in his immediate review of the *Origin*, he was happy to go public on the matter. "Extinguished theologians lie about the cradle of every science as the strangled snakes beside that of Hercules; and history records that whenever science and orthodoxy have been fairly opposed, the latter has been forced to retire from the lists, bleeding and crushed, if not annihilated; scotched, if not slain" (Huxley 1860, 52).

Huxley's readers, friends and foes, saw what he was about, starting with the fact that he called his essays "Lay Sermons": "He has the moral earnestness, the volitional energy, the absolute conviction in his own opinions, the desire and determination to impress them upon all mankind, which are the essential character- istics of the Puritan character. His whole temper and spirit is es- sentially dogmatic of the Presbyterian or Independent type, and he might fairly be described as a Roundhead who had lost his faith." He shows all the signs: "the hortatory passages, the solemn per- sonal experiences, the heart-searchings and personal appeals that are found in Puritan literature" (Baynes 1873, 502).

Huxley inspired others. In the bestselling novel *Robert Elsmere* by Mrs. Humphrey Ward (Mary Augusta Ward), the hero—an Anglican vicar—loses his faith, in part through reading Darwin (on earthworms!). Before our cleric succumbs at too early an age—he mistook something far more serious for a passing instance of "parson's throat"—he devotes his life to secular Sunday schools in London's East End, something spe- cifically advocated by Huxley in those lay sermons. Not just in fiction. Huxley's grandson, the evolutionary biologist Julian Huxley, incidentally also the nephew of Mrs. Ward, was even keener to make a religion out of his science. He wrote a book called *Religion without Revelation*. In the pattern of the older man, grandfather Thomas Henry, Julian did not want to rid the world of religion. He wanted to change it for secular purposes. God must go, but what remains of religion is vital: "if, finally, there be no reason for ascribing personality or pure spirituality to this God, but every reason against it; then religion becomes a natural and vital part of human existence, not a thing apart; a false dualism is overthrown; and the pursuit of the religious life is seen to resemble the pursuit of a scientific truth or artistic expression, as the highest of human activities" (Huxley 1927, 53–54). We need religion. Not the traditional religion. No prizes for guessing what the future author of the classic *Evolution: The Modern Synthesis* (Huxley 1942) had in mind.

Prominent among today's evolutionists, Edward O. Wilson likewise sets about making a religion from his science. It is materialistic, or at least naturalistic, as it presents "the human mind with an alternative mythology that until now has always, point for point in zones of conflict, defeated traditional religion." Helpfully, Wilson tells us: "Its narrative form is the epic: the evolution of the universe from the big bang of fifteen billion years ago through the origin of the elements and celestial bodies to the beginnings of life on earth. The evolutionary epic is mythology in the sense that the laws it adduces here and now are believed but can never be definitively proved to form a cause-and-effect continuum from physics to the social sciences, from this world to all other worlds in the visible universe, and backward through time to the beginning of the universe." After this, we are hardly surprised to learn that "If this interpretation is correct, the final decisive edge enjoyed by scientific naturalism will come from its capacity to explain traditional religion, its chief competition, as a wholly material phenomenon. Theology is not likely to survive as an independent intellectual discipline" (Wilson 1978, 192).

Progress versus Providence

Take up now the need to identify the added ingredient that is to make all this possible. If we are to have a religion—secular or otherwise—we need an underlying metaphysic to hold it together. To make a picture. To confer meaning. This will be a kind of root metaphor. In the case of Christianity, although there are variations, we find our metaphysic, our root, in the idea of Providence. A Creator God, on whom we are totally dependent, who so loved us that for us He made the supreme sacrifice. An idea well expressed by the Congregationalist minister Isaac Watts in his great hymn of 1707:

> When I survey the wondrous cross
> On which the Prince of glory died,
> My richest gain I count but loss,
> And pour contempt on all my pride.

What does evolution have to offer in its stead? If not Providence, then what? Already, in looking at Erasmus Darwin, we have had strong intimations about what this might be. It is the idea that Darwin expressed at the end of the poetic passage quoted in chapter 1: "Imperious man, who rules the bestial crowd. . . . Arose from rudiments of form and sense, An embryon point, or microscopic ens!" (Darwin 1803, 1, 11, 314). And then he tied it in with a more general philosophy of progress, telling us that the idea of organic progressive evolution "is analogous to the improving excellence observable in every part of the creation;

such as the progressive increase of the wisdom and happiness of its inhabitants" (Darwin 1794–1796, 2, 247–248). This idea of progress—things getting better and better—in the cultural world and then reflecting into the biological world—thus conferring meaning as strongly as Christian Providence—continued to get major and sympathetic attention in the years before Charles Darwin. It was the underlying theme of a very popular, pre-Darwinian, evolutionary tome, *Vestiges of the Natural History of Creation* (first published in 1844). This added passage to the fifth edition makes the point explicitly:

> A progression resembling development may be traced in human nature, both in the individual and in large groups of men. . . . Now all of this is in conformity with what we have seen of the progress of organic creation. It seems but the minute hand of a watch, of which the hour hand is the transition from species to species. Knowing what we do of that latter transition, the possibility of a decided and general retrogression of the highest species towards a meaner type is scarce admissible, but a forward movement seems anything but unlikely. (Chambers 1846, 400–402)

This is a major reason why the work so upset many leading scientists like Adam Sedgwick, William Whewell, and David Brewster (Ruse 1979a). They were sincere Christians, often under pressure from more conservative co-religionists who feared that science was inimical to Christianity. They could and did brush off modern-day-creationist-like objections, based on six days of Creation and universal floods and so forth. Progress was another matter. Here was proof that evolution, through progress, claiming to be genuine science, went negatively to the heart of conventional Christianity. Both Providence and progress are aimed at a good ending, but whereas for Christianity it all comes down to God, for the progressionist it is all a matter of rolling up our sleeves and doing it ourselves. (This controversy, led by men who were the mentors of Charles Darwin, is almost certainly the reason for the delay in publishing the theory of evolution through natural selection. Darwin waited until the old guard was no longer so active and influential.)

Then, in the 1850s, came the indefectible Herbert Spencer. For him, progress was everything. From an article two years before the *Origin*:

> Now we propose in the first place to show, that this law of organic progress is the law of all progress. Whether it be in the development of the Earth, in the development of Life upon its surface, in the development of Society, of Government, of Manufactures, of Commerce, of Language, Literature, Science, Art, this same evolution of the simple into the complex, through successive differentiations, holds throughout. From the earliest traceable cosmical changes down to the latest results of civilization, we shall find that the transformation of the homogeneous into the heterogeneous is that in which Progress essentially consists. (Spencer 1857, 245)

As we rush down through the history of Darwinian theorizing, we find that progress continues to ride high in evolutionary circles. With acknowledgment to Spencer, Julian Huxley defined evolutionary progress as "increased control over and independence of the environment" (1942, 545). Thus defined, humans come out on top. Although, remembering the wisdom of Spider-Man's Uncle Ben, with great power comes great responsibility. "The future of progressive evolution is the future of man. The future of man if it is to be a progress and not merely a standstill or a degeneration, must be guided by a deliberate purpose. And the human purpose can only be formulated in terms of the new attributes achieved by life in becoming human" (577). One thing was that Huxley knew the enemy when he saw it. Thanks to its underlying metaphysic, Christianity leads to moral intellectual and physical laziness. "Divine Providence is an excuse for the poor whom we have always with us; for the human improvidence which produces whole broods of children without reflection or case as to how they shall live; for not taking action when we are lazy; or, more rarely, for justifying the action we do take when we are energetic. From the point of view of the future destiny of man, the present is the time of clash between the idea of Providentialism and the idea of humanism—human control by human effort in accordance with human ideals" (Huxley 1927, 18). One should say that Huxley put his money where his mouth was. He was the first director general of the new United Nations Educational, Scientific and Cultural Organization (UNESCO). It was he who insisted on the S for Science. Like the early Christians, he suffered for his faith. Because of a provocative pamphlet (Huxley 1948) he wrote about his vision for UNESCO—a vision entirely infused with progress—he was denied a full four-year term.

Wilson is open in his fervent belief in biological progress: "The overall average across the history of life has moved from the simple and few to the more complex and numerous. During the past billion years, animals as a whole evolved upward in body size, feeding and defensive techniques, brain and behavioral complexity, social organization, and precision of environmental control—in each case farther from the nonliving state than their simpler antecedents did." Adding: "Progress, then, is a property of the evolution of life as a whole by almost any conceivable intuitive standard, including the acquisition of goals and intentions in the behavior of animals" (1992, 187).

Wilson, Southern born and bred and saved from his sins as a teenager, has moved on from his early years. Out of respect and affection, compared to the Huxleys, he is a lot less hostile toward Christianity. Candor, nevertheless, forces him to admit that his heart now belongs to a rival. "I see no way to avoid the fundamental differences in our respective worldviews." On the one hand, Christianity: "You are a literalist interpreter of Christian Holy Scripture. You reject the conclusion of science that mankind evolved from lower forms. You believe that each person's soul is immortal, making this planet a way station to a second, eternal life. Salvation is assured those who are redeemed in Christ." On the other hand, Darwinism: "I am a secular humanist. I think existence is what we make of it as individuals. There is no guarantee of life after death, and heaven and hell are what we create for ourselves, on this planet. There is no other home. Humanity originated here by evolution from lower forms over millions of years." We are fancy apes who have adapted rather well to life here on earth. And this means that spiritual explanations and understandings are otiose. The same is true of behavior: "Ethics is the code of behavior we share on the basis of reason, law, honor, and an inborn sense of decency, even as some ascribe it to God's will" (Wilson 2006, 3–4).

Is there Meaning?

Evolutionists do not have identical thoughts about progress. T. H. Huxley rather lost faith toward the end of his life; the horrendous social conditions in Victorian cities and like phenomena shook his thinking. He was not against all thoughts of progress, but they were tempered (Huxley 1893). Christians likewise do not have identical thoughts about Providence. Think of the difference on this topic between, let us say, the Roman Catholics and the Jehovah's Witnesses. And that is to omit the Calvinists! The important thing is that, given the notion of Providence, meaning falls at once into place. We are to do our duty by God—praising Him and following His orders. On its own, our efforts can never be enough. But God through His love and His mercy forgives us. "In my Father's house are many mansions: if it were not so, I would have told you. I go to prepare a place for you" (John 14:2). That is meaning or Meaning—all that we have and all that we could ever desire. What then about meaning or Meaning for the evolutionary progressionist? Although he was no evolutionist and although he was in respects still deeply indebted to Providential thinking—as were many, including Charles Darwin—the visionary poet and painter William Blake caught the spirit of things. In the poem now called "Jerusalem," he refers to Jesus's supposed visit to England

and to the grim aspects of the Industrial Revolution. We are urged to action:

> And did those feet in ancient time
> Walk upon England's mountains green:
> And was the holy Lamb of God,
> On England's pleasant pastures seen!
>
> And did the Countenance Divine,
> Shine forth upon our clouded hills?
> And was Jerusalem builded here,
> Among these dark Satanic Mills?
>
> Bring me my Bow of burning gold;
> Bring me my Arrows of desire:
> Bring me my Spear: O clouds unfold!
> Bring me my Chariot of fire!
>
> I will not cease from Mental Fight,
> Nor shall my Sword sleep in my hand:
> Till we have built Jerusalem,
> In England's green & pleasant Land.

Turned into a hymn and set to the music of Hubert Parry, the words are always sung at the closing of the annual conference of the British Labor Party. I am not surprised. British socialism owes more to Methodism than to Marxism, and more to the increasingly secular optimism of the nineteenth century than to Methodism.

Meaning for the evolutionist is found in the upward rise of the history of life—monad to man. We humans are in some objective sense the winners, the top of the tree, of more value than other organisms. This is a function of many things, but our minds, our consciousness, our intelligence, are the all-important factors. We are in some sense more complex than other organisms. This complexity, in some way, plays itself out by making us thinking beings with an ability to understand our world and with our own powers of choice, of deciding between good and evil. This readily translates into prescriptions. In the biological world, we are to keep up the evolutionary process, at least not letting it decline and perhaps helping it ever upward. In the social realm, for remember that biological progress is a child of cultural progress, we are to make for a better society for one and for all. Thomas Henry Huxley did not sit on the London School Board by chance. Julian Huxley did not become director general of UNESCO by chance. Edward O. Wilson did not win numerous awards, like the Tyler Prize for Environmental Achievement, by chance. The endpoint is still in this world, but it has great value in itself and gives meaning to the lives of those who strive to realize it.

Why Progress?

I shall have more to say in a moment about the rules of behavior that this kind of thinking generates, both their nature (substantive or normative ethics) and their standing (metaethics). First, however, I want to turn to a question that is becoming increasingly pressing. We have been talking about evolutionary progress. How does natural selection fit into all of this? To be honest, thus far, not much (Ruse 1996). Based on some half-baked readings of physics—a lot of his thinking was based on half-baked readings—Spencer saw life naturally in a kind of equilibrium (Richards 1987). Then something disturbs this equilibrium, and there is turmoil and chaos. Finally, equilibrium is reachieved but at a higher point. Interestingly, into this potpourri of thinking—"dynamic equilibrium"—Spencer introduced the Malthusian population explosion leading to struggle. For Spencer, however, the struggle led not so much to failure and selection as to striving to succeed, with the winners passing on their better attributes through a Lamarckian process. Apparently, sharing his fellow Victorians' belief that an organism can produce only a limited amount of vital bodily fluid, Spencer argued that either it can flow out between the loins into making many offspring, or it can head up to the brain and make for ever greater intelligence. The consequence is that with the more intelligent producing ever fewer children, the Malthusian struggle slows down and stops. Spencer himself was so far advanced that he was a lifelong bachelor and had no offspring at all.

Although Spencer was notoriously stingy in admitting influences, his belief in a kind of progressive force came from the German Romantics, above all from Friedrich Schelling, whose ideas were taken up wholesale (Ruse 2013). It didn't hurt that the poet and essayist Samuel Coleridge had plagiarized Schelling and translated key passages into English. Be this as it may, Spencer's influence lived on. Julian Huxley was open about his debt, and Wilson likewise acknowledges an influence. His intellectual grandfather at Harvard, the supervisor of his supervisor, was W. H. Wheeler, a fellow ant specialist and Spencer groupie. In Wilson's lab, there was a picture of Spencer on the wall hung more prominently than that of Darwin. Admittedly, neither as prominent as the picture of Wilson receiving the National Medal of Science from President Jimmy Carter. Hero worship has its limits.

What then of the prospect of getting progress out of a process that, if not entirely selection-driven, certainly has natural selection as a significant causal factor? It is instructive to begin with Charles Darwin himself. As one raised an Anglican, he would be looking for the big picture to make sense of all, and at first this clearly would be Providence. As his faith in the Christian Creator faded, his need for a big picture was no less, and progress slides readily in to take the place of Providence. We still have a desirable end and the need for us to strive toward it. It is just that now we do it all by ourselves, even if there is an Unmoved Mover behind it all. For Darwin particularly, the move to progress was natural, for he was a child—actually grandchild—of one of the greatest successes of the Industrial Revolution. His maternal grandfather was Josiah Wedgwood the potter. There was great wealth in the Darwin family, something added to when Charles married his first cousin, Emma Wedgwood, also a grandchild of old Josiah.

Darwin came naturally to progress, but he knew also that there is a worm in the bud. Most commercial enterprises fail. The number of bankruptcies far exceed the number of successes. Progress in the Darwinian world can never be the easy upward movement, a kind of naturalized Providence, where all will come right, no matter. He wrestled with this in a private notebook:

The enormous number of animals in the world depends of their varied structure & complexity.—hence as the forms became complicated, they opened fresh means of adding to their complexity.—but yet there is no necessary tendency in the simple animals to become complicated although all perhaps will have done so from the new relations caused by the advancing complexity of others.—It may be said, why should there not be at any time as many species tending to dis-development (some probably always have done so, as the simplest fish), my answer is because, if we begin with the simplest forms & suppose them to have changed, their very changes tend to give rise to others. (in Barrett et al. 1987, E 95–97)

Complexity gives rise to new opportunities, and increasingly Darwin saw that natural selection can aid organisms to take advantage of these.

Spencerian-type thinking before its time. One is arguing that progress will necessarily occur because nature has an inbuilt drive to complexity. Increasingly, as grew his confidence in natural selection, Darwin dropped this kind of thinking. But not the belief in progress of some kind or another. The *Origin* does not discuss humans in detail. It still makes clear Darwin's belief in progress: "The inhabitants of each successive period in the world's history have beaten their predecessors in the race for life, and are, in so far, higher in the scale of nature; and this may account for that vague yet ill-defined sentiment, felt by many palaeontologists, that organisation on the whole has progressed" (Darwin 1859, 345). The closing passage of the book repeats that sentiment. The question is how selection brings this about. Through competition, obviously, but precisely how?

Even Darwin's supporters made clear the difficulties. On the one hand, natural selection is relativistic. What works in one situation might not work in another. There is no one, unique, always-winning adaptation. Paleontologist Jack Sepkoski makes the point colorfully: "I see intelligence as just one of a variety of adaptations among tetrapods for survival. Running fast in a herd while being as dumb as shit, I think, is a very good adaptation for survival" (quoted in Ruse 1996, 486). Cow power rules supreme! On the other hand, and something that really worried the Harvard botanist Asa Gray (1876), Darwin always insisted (as do today's Darwinians) that the raw blocks of evolution, the new variations—known today as mutations, produced by changes in the units of heredity, the genes—are random. This does not mean uncaused—Darwin thought variations might be caused by external phenomena impinging on the reproductive organs—but it does mean not occurring to need. If a new predator arrives on the scene, it might be of advantage to turn dark for better camouflage. As like as not, any variations you get could be pink or green. Gray thought this meant that real advance was impossible, and so he supposed guided variations, which fit in nicely with his strong Presbyterian faith. And which was anathema to Darwin, who, deist or agnostic, wanted to keep God at a distance.

By the third edition of the *Origin*, in 1861, Darwin was confident that he had the answer. Progress thanks to selection, rather than progress despite selection. He invoked what today's Darwinians call "arms races"—lines compete against each other, and one line gets better and better. Eventually this comparative improvement is translated into some form of absolute improvement, *Homo sapiens*:

> If we look at the differentiation and specialisation of the several organs of each being when adult (and this will include the advancement of the brain for intellectual purposes) as the best standard of highness of organisation, natural selection clearly leads towards highness; for all physiologists admit that the specialisation of organs, inasmuch as they perform in this state their functions better, is an advantage to each being; and hence the accumulation of variations tending towards specialisation is within the scope of natural selection. (1861, 134)

Natural selection cannot guarantee anything, but everything is probably going to be just fine. The great success of capitalism and British industry leads the way. When it came to humans, Darwin was confident that selection didn't produce just our species but the better instances of our species. To a doubting correspondent toward the end of his life, he wrote: "I could show fight on natural selection having done and doing more for the progress of civilisation than you seem inclined to admit. Remember what risks the nations of Europe ran, not so many centuries ago of being overwhelmed by the Turks, and how ridiculous such an idea now is. The more civilised so-called Caucasian races have beaten the Turkish hollow in the struggle for existence. Looking to the world at no very distant date, what an endless number of the lower races will have been eliminated by the higher civilised races throughout the world" (Darwin Correspondence Project, letter 13230, to William Graham, July 3, 1881). Darwin wasn't always quite this chauvinistic. In *The Descent of Man*, nearly ten years earlier, he had credited the success of Europeans over other races to their greater immunity to the diseases that they passed on to others. They survived, the other races didn't.

Today's Evolutionists and Progress

Arms races! Today, Dawkins stands in this tradition. "Directionalist common sense surely wins on the very long time scale: once there was only blue-green slime and now there are sharp-eyed metazoa" (Dawkins and Krebs 1979, 508). The key lies in arms races. Having embraced computer technology early and enthusiastically, Dawkins slides easily into noting that, more and more, today's arms races rely on computer technology rather than brute power. In the animal world, Dawkins finds this translated into ever bigger and more efficient brains. Oh, what a surprise! We humans are the winners! Dawkins invokes a notion known as an animal's EQ, standing for "encephalization quotient" (Jerison 1973). This is a kind of cross-species measure of IQ that factors in the brain power needed simply to get an organism to function (whales require much bigger brains than shrews because they need more computing power to get their bigger bodies to function) and then scales according to the surplus left over. Dawkins writes: "The fact that humans have an EQ of 7 and hippos an EQ of 0.3 may not literally mean that humans are 23 times as clever as hippos! But the EQ as measured is probably telling us something about how much 'computing power' an animal probably has in its head, over and above the irreducible amount of computing power needed for the routine running of its large or small body" (1986, 189).

Leave critical comment for a moment. Turn to another approach that claims to have found a selection-driven progressive process. Paleontologist Simon Conway Morris (2003)—a Christian but seeking an entirely natural explanation—argues that only certain areas of what we might call "morphological space" are welcoming to life forms. The center of the sun would not be, for instance. This constrains the course of evolution. Again and again, organisms take the same route into a pre-existing niche. The sabertooth-tiger-like organisms are a nice example, where the North American placental mammals (real cats) were matched right down the line by South American marsupials (thylacosmilids). There existed a niche for organisms that were predators, with cat-like abilities and shearing/stabbing-like weapons. Darwinian selection found more than one way to enter it—from the placental side and from the marsupial side. It was a question not of beating out others but of finding pathways that others had not found.

Conway Morris argues that, given the ubiquity of convergence, the historical course of nature is not random but strongly selection-constrained along certain pathways and to certain destinations. Most particularly, some kind of intelligent being was bound to emerge. After all, our own very existence shows that a kind of cultural adaptive niche exists—a niche that prizes intelligence and social abilities. "If brains can get big independently and provide a neural machine capable of handling a highly complex environment, then perhaps there are other parallels, other convergences that drive some groups towards complexity." Continuing: "We may be unique, but paradoxically those properties that define our uniqueness can still be inherent in the evolutionary process. In other words, if we humans had not evolved then something more-or-less identical would have emerged sooner or later" (Conway Morris 2003, 196).

Finally, for completeness in this mini-survey of today's thinking about evolution and progress, let me circle back to a solution more Spencerian than Darwinian, although you can see it contains elements of the ideas that fascinated Darwin long ago. Duke University colleagues paleontologist Daniel McShea and philosopher Robert Brandon promote what they proudly call "Biology's First Law." Termed the "zero-force evolutionary law," or ZFEL, its general formulation states: "In any evolutionary system in which there is variation and heredity, there is a tendency for diversity and complexity to increase, one that is always present but that may be opposed or augmented by natural selection, other forces, or constraints acting on diversity or complexity" (2010, 4). McShea and Brandon see this law as something with the status in evolutionary biology of Newton's First Law of Motion. It is a kind of background condition of stability—perhaps better, continuity—against which other factors can operate.

The authors are fairly (let us say) generic in their understandings of complexity and diversity—number of parts, number of kinds. Whatever, the claims to be made are grandiose if familiar. Given the natural tendency of life toward complexity—parts tend to be added on—this generates new organic variations and hence types. One thus gets a version of what the theoretical biologist Stuart Kauffman (1993) has called, "order for free." It is not always obvious whether the claim is that adaptation is created in this way or if adaptation is now irrelevant.

Probably more the former: "we raise the possibility that complex adaptive structures arise spontaneously in organisms with excess part types. One could call this self-organization. But it is more accurately described as the consequence of the explosion of combinatorial possibilities that naturally accompanies the interaction of a large diversity of arbitrary part types" (McShea and Brandon 2010, 124). Their feeling is that selection is mainly a negative force, cleaning up after the really creative work has been done. Naturally, all of this is going to lead to intelligence. "The scope we claim for the ZFEL is immodestly large. The claim is that the ZFEL tendency is and has been present in the background, pushing diversity and complexity upwards, in all populations, in all taxa, in all organisms, on all timescales, over the entire history of life, here on Earth and everywhere" (124).

Skepticism

What is there to say about all this progress making? Basically, the more Darwinian you get, the less likely progress is to emerge, and the more likely progress is to emerge, the less Darwinian you get. One can say for certain that not all (professional) Darwinian evolutionists have been that enthusiastic. Many are lukewarm toward the claims made in the name of arms races (Ruse 1996). There is undoubtedly some empirical evidence for them. For instance, as predators get better at boring into shells, the owners of those shells get better at producing ever stronger and thicker protection.

However, the evidence is not uniformly positive. Fossils, for instance, do not show unambiguously that prey and predators have become ever faster. And even if arms races are ubiquitous, it does not follow that intelligence will always emerge. Having high intelligence means having large brains, and having large brains means having ready access to large chunks of protein, the bodies of other animals. There were no clever vegans in the Pleistocene. Being clever, however, isn't everything. Sometimes—as cows and horses demonstrate—it is just easier to get your food in other ways, especially if you are living on grassy savannahs. Remember the cautionary words of Sepkoski.

Turning to the second suggestion, even if it exists, why should we or anyone else necessarily or even probably enter the culture niche? Life is full of missed opportunities. Maybe most times evolution would have gone other ways and avoided culture entirely. Warthogs rule supreme. Many wonder if it is right to think that niches are just waiting out there, ready to be conquered and entered (Lewontin 2000). Don't organisms create niches as much as find them? Until vertebrates like us humans came along, there was hardly a niche for head lice, for instance. Should we expect that there was a niche for culture, just waiting there, like dry land or the open air? Perhaps there are other niches not yet invented. We cannot imagine something other than consciousness, but that could be our problem, not a fault of objective reality. For all their talk about analogy, Christians tend to think that their God can get up to some clever tricks, way beyond their ken. Perhaps these are not all supernatural abilities but simply abilities that were omitted from our evolution. Perhaps, far from being the best, we are a short side path and very limited in the true scheme of things. No more than in the case of arms races do we get much guarantee of either human emergence or a sense that we are in some way superior and for this reason we won.

ZFEL? One can only say that if you believe in this, then you are ready for the tooth fairy. Ideas like this seem to appeal to, and only to, those who spend their days in front of computers. If they spent less time running simulated scenarios and more time looking at the real world, they would think otherwise. Blind law leads to blind results, and supposing that laws are not really blind is to suppose, like Gray, that there is some kind of intelligence behind it all. Popular these days among philosophers is an Aristotelian force, an Unmoved Mover, to which everything strives teleologically. I just don't see the evidence for it, I just don't. In my world, Murphy's Law rules supreme. If it can go wrong, it will go wrong. I am not sure that either arms races or niches did do the job, but I can see how they could and how natural selection was vital. Suggestions like ZFEL strike me as trying to slip in some kind of (German Romantic) teleological force upward to humans. I am not necessarily saying this is wrong—Aristotle did it, and anyone less shifty than he it would be hard to imagine—but it is being done by sleight of hand. Some, like the philosopher Thomas Nagel (2012), are quite open about doing this. He truly does believe in, or in the equivalent of, an Aristotelian Unmoved Mover. To which I can only say that I have not seen much evidence of it in my life or the lives of others. Would that Heinrich Himmler had had some intimation of it as he formulated the Final Solution.

What about Meaning?

Hold, for a moment, final judgments about natural selection and progress, and turn to the final part of the discussion—meaning. The argument is straightforward. It has certainly seemed so to enthusiasts from Spencer to Wilson. Evolution is progress, which means that it is a value-laden phenomenon. Over time, value increases. Humans are the endpoint, which means we are the most valuable species here on earth. Humans over warthogs over reptiles over fish over bacteria. Hence, meaning comes through humans, cherishing us, protecting us, keeping us from decline—biological or cultural—and if possible, helping us to advance even further—biological and cultural. Well known are various eugenical movements that have promoted biological improvement or, at the very least, selective culling to avoid decline. Interestingly, perhaps because they know that this is far more difficult than most enthusiasts presume, they incline more toward cultural processes for improvement. Spencer is probably mixing both:

> we have been led to see that Ethics has for its subject-matter, that form which universal conduct assumes during the last stages of its evolution. We have also concluded that these last stages in the evolution of conduct are those displayed by the highest type of being when he is forced, by increase of numbers, to live more and more in presence of his fellows. And there has followed the corollary that conduct gains ethical sanction in proportion as the activities, becoming less and less militant and more and more industrial, are such as do not

necessitate mutual injury or hindrance, but consist with, and are furthered by, co-operation and mutual aid. (1879, 21)

Julian Huxley had the same message, in an updated form. He was interested in eugenics, but his heart was in cultural improvements. He was much taken by large-scale technological achievements like the Tennessee Valley Authority (TVA), which brought power right across the South, until then a Third World country indeed. Huxley had to temper his enthusiasm, given that this was also the time of Adolf Hitler, but the message is clear:

All claims that the State has an intrinsically higher value than the individual are false. They turn out, on closer scrutiny, to be rationalizations or myths aimed at securing greater power or privilege for a limited group which controls the machinery of the State.

On the other hand the individual is meaningless in isolation, and the possibilities of development and self-realization open to him are conditioned and limited by the nature of the social organization. The individual thus has duties and responsibilities as well as rights and privileges, or if you prefer it, finds certain outlets and satisfactions (such as devotion to a cause, or participation in a joint enterprise) only in relation to the type of society in which he lives. (1943, 138–139)

Wilson, writing today at the time of environmental crises like the rape and destruction of the Brazilian rainforests, shows his moral concerns by arguing that humans have developed in symbiotic relationship with the rest of nature and that we can survive only in a world of biological diversity. We have a natural inclination to what Wilson calls "biophilia" (1984). A world of plastic would kill, literally as well as metaphorically. For this reason, we must promote the preservation of such diversity. As Wilson writes in a more recent book, *The Future of Life*: "a sense of genetic unity, kinship, and deep history are among the values that bond us to the living environment. They are survival mechanisms for us and our species. To conserve biological diversity is an investment in immortality" (2002, 133). More recently, at the age of eighty-six, he has written: "Laid before us are new options scarcely dreamed of in earlier ages." Hitting right on our topic: "If the heuristic and analytic power of science can be joined with the introspective creativity of the humanities, human existence will rise to an infinitely more productive and interesting meaning" (2014, 187).

One is surely sympathetic to the prescriptions (in the language of the philosophers, the substantive ethics) of these Darwinian ethicists. Despite a reputation as an exponent of extreme laissez-faire, as the above-quoted passage suggests, Spencer was concerned about society and highly critical of empire building and militarism. Who could deny that the TVA was a very good thing? No argument is needed for the worth of Wilson's concerns. The trouble is that no one is very Darwinian in their thinking about progress. No big surprise. Trying to get meaning out of Darwin's theory (going for metaethical justification) is a bit like trying to square the circle. It can't be done. Certainly, it cannot be done by progress. The Darwinian mechanism of natural selection, working on random variations, is for progress what Daniel Dennett (1995) calls a "universal acid." Natural selection is relative; variations are random. All the enthusiasts we have just been looking at were less

Darwinian and more Spencerian and in the Romantic tradition. By their own admissions, even though, interestingly, it was Julian Huxley who first developed the notion of a biological arms race, using as an analogy the pre-World War I naval race between Britain and Germany. Not to worry. When it came to human evolution, Huxley stepped sharply into (non-Darwinian) line, arguing that it could only have taken the course that it did and acknowledging the great influence of the French philosopher Henri Bergson. It figures, because Bergson was a vitalist, seeing forces pushing upward—"in the last analysis, man might be considered the reason for the existence of the entire organization of life on our planet" (1911, 185). Directly or indirectly through Spencer (whose influence is acknowledged by both Bergson and Huxley), the Romantic tradition is the foundation.

The Naturalistic Fallacy

The knife is in. If you are a Darwinian, no progress; if you have progress, then you are not a Darwinian. It would be churlish not to note that lurking behind this discussion is what philosophers, following G. E. Moore (1903), called the "naturalistic fallacy," trying illicitly to get non-natural phenomena—morality, meaning—from natural phenomena—the path of evolution. It is that all-important second item identified in chapter 2 as questions not answered by science. Value from fact. Preceding Moore, this is a violation of what has come to be known as Hume's Law:

In every system of morality, which I have hitherto met with, I have always remark'd, that the author proceeds for some time in the ordinary way of reasoning, and establishes the being of a God, or makes observations concerning human affairs; when of a sudden I am surpriz'd to find, that instead of the usual copulations of propositions, is, and is not, I meet with no proposition that is not connected with an ought, or an ought not. This change is imperceptible; but is, however, of the last consequence. For as this ought, or ought not, expresses some new relation or affirmation, 'tis necessary that it shou'd be observ'd and explain'd; and at the same time that a reason should be given, for what seems altogether inconceivable, how this new relation can be a deduction from others, which are entirely different from it. (Hume 1739–1740, III, I, 1)

You may agree that humans are superior to warthogs. I do, too, although after a trip to Zimbabwe, I do confess a sneaking affection for them. The point is that you don't get this out of evolutionary theory. The Sepkoski objection again. I can think of many situations in which warthogs do very nicely—in Zimbabwe for a start—and I cannot honestly always say the same of humans—in Zimbabwe for a second. I value intelligence and kindness and much more, but evolutionary theory—Darwinian evolutionary theory, in particular—is rather standoffish about them. Carnivores make their livings out of killing other animals—usually with much pain. Intelligence, as we have seen, is a mixed blessing. What is good for me is very much not necessarily good for you.

The fact is, however, the people we are talking about don't accept Hume's Law or the naturalistic fallacy. After writing a paper on evolution and ethics with Wilson, I can attest personally that he simply doesn't agree that one cannot get value out of facts (Ruse and Wilson 1986). To the objection that value language doesn't occur in evolutionary thinking—other than relative values like "fitter than"—he responds that in deductions, language often changes. Gas theory speaks of molecules at one level and gases at another. Facts at one level, values at another. Like many scientists venturing into the world of philosophy, Wilson wears hobnailed boots. But there are those in the philosophical community who are empathetic. Philip Kitcher, for instance, a keeper of the sacred flame of Darwin's thinking against heretics—including Wilson!—speaks in terms of "humanism." He means, presumably, "Humans one, Warthogs zero." He writes movingly, in *Life after Faith: The Case for Secular Humanism* (2014), of seeing the ever-greater worth of the "ethical project." Although Kitcher is critical of raw spiritual religion, as it were, he writes sympathetically about its possibilities: "We need to make secular humanism responsive to our deepest impulses and needs, or to find, if you like, a cosmopolitan version of spiritual religion that will not collapse back into parochial supernaturalism" (2007, 162).

What about the is/ought barrier? Kitcher, a Darwinian, is also a pragmatist, or Pragmatist, much in the tradition of John Dewey. He brushes the whole debate to one side, implying that it truly is a non-problem that takes our attention from the real issues. To ask for more, he tells us—to ask for a proof that you can go from is to ought—is "bizarre."

> Human beings do not wander around the world, gathering facts and nothing but facts, until one day, they launch themselves into ethical life by some specially adroit piece of inference. From our earliest stages as thinking beings, we are immersed in a mixture of factual beliefs and value judgments, transmitted to us by our elders—we are "always already" ethical (Kitcher 2014, 53).

There is an important truth here. You must take some kind of position like this toward Hume's Law—simply ignore it or brush it aside as irrelevant—or you are impaled on it.

Perhaps Kitcher is right. At a fundamental level none of this really matters, because whatever we say, we think and act differently. A year or two back, I spent a semester in South Africa, in Stellenbosch at an institute affiliated with the Afrikaans university there. Even if it were not in the heart of the wine country, with opportunities to spend all day and every day at a different winery sampling their products, it is truly in the running as the loveliest place on earth—vineyards running up the sides of mountains of breathtaking beauty. There was talk of a mining company coming in and taking off the top of one of the mountains. The immediate reaction was: This is rape. And if that is not to anthropomorphize the world and give it value, I do not know what is. Was it just a metaphor? I am not sure. In some sense, perhaps that valley really did have value in its own right.

And the World Said?

Am I not giving the whole game away, allowing that biology does show that the world has value and thus we have an answer to meaning, both secular and objective? Not really. Even if you do think the world has value, without an added ingredient, there is nothing to say that a human is of more worth than a tree or a mountaintop. In fact, the whole point is that you are not into ordering. Humans have value. Trees have value. Mountaintops have value. There is no progress there unless you start to infuse the discussion with some kind of value-seeking consciousness. Darwin doesn't help. Which takes us right back to Romanticism, a worldview that has many of the problems identified in chapter 2 as making Christianity implausible (Richards 2003). If the Christian God is out, so also are world spirits and the like. In the thought of someone like G. W. F. Hegel, I am not sure that they are all that different. Even though one might pull back from some of the political views he abstracted from it, Hegel's idea of Absolute Spirit seems in tune with a venerable mystical approach to the Christian Godhead (Taylor 1975). The Quaker in me responds sympathetically to this understanding.

This does not deny that, in the end, my reactions to Darwinism as religion are guarded, distinctly more negative than positive. Can you extract objective meaning from the theory of evolution through natural selection? Some serious scientists think you can, although often (always!) one feels the real motivation is a kind of neo-Romantic view of nature and its changes. Essentially non-Darwinian, although (as in the case of people like Dawkins) using natural selection to paper over the gaps and problems. Are humans really so special? Well, yes, we usually think they are, but if this conviction comes from or is read into the science is another matter. Can the world have objective meaning in its own right? Some think it can. Some think it can't. Some are on the fence. Why not then leave our discussion here? Not entirely satisfactory, but with arguments that convince some and that might get yet stronger with more empirical work—about arms races and about niches, for example. Ultimately, however, the Humean skeptic within will out. I worry about this whole strategy and if it is a bit of a cop-out. Regular religion cannot do the job, so make up a natural-istic religion of your own. Neither approach takes the Darwinian Revolution quite as seriously as it should. The arguments of the previous chapters are decisive. Darwinian thinking takes meaning out of the world. Hardy was right:

> —Crass Casualty obstructs the sun and rain,
> And dicing Time for gladness casts a moan. . . .
> These purblind Doomsters had as readily strown
> Blisses about my pilgrimage as pain.
>
> ("Hap")

We are on our own! We are on our own! God and the soul don't protect us. Nor does it help to turn nature into a meaning gen-erator. We are on our own! Is this just a counsel of despair? Was Shakespeare right?

> Life's but a walking shadow, a poor player
> That struts and frets his hour upon the stage

And then is heard no more: it is a tale
Told by an idiot, full of sound and fury,
Signifying nothing.

(*Macbeth*)

I can only ask that you don't give up yet. We have not thus far tried a subjective naturalistic approach, one that appreciates and even uses Darwinian theory. Let us see where that leads us.

4 | DARWINIAN EXISTENTIALISM

For meaning, we must search within ourselves, and we must do it in the light of Darwinian evolutionary theory, both the fact that we are animals and that this is a bleak world indeed. "We must search within ourselves"? "This is a bleak world indeed"? Even for the philosophy undergraduate, this at once strikes a chord. Jean-Paul Sartre! He makes the point about the alienation from God:

> Existentialism is not so much an atheism in the sense that it would exhaust itself attempting to demonstrate the non-existence of God; rather, it affirms that even if God were to exist, it would make no difference—that is our point of view. It is not that we believe that God exists, but we think that the real problem is not one of his existence; what man needs is to rediscover himself and to comprehend that nothing can save him from himself, not even valid proof of the existence of God. (Sartre 1948, 56, based on a lecture given in 1945)

He tries to explain what this means for humankind:

> My atheist existentialism ... declares that God does not exist, yet there is still a being in whom existence precedes essence, a being which exists before being defined by any concept, and this being is man or, as Heidegger puts it, human reality.
>
> That means that man first exists, encounters himself and emerges in the world, to be defined afterwards. Thus, there is no human nature, since there is no God to conceive it. It is man who conceives himself, who propels himself towards existence. Man becomes nothing other than what is actually done, not what he will want to be. (27–28)

This is a bleak world indeed! We must search within ourselves! "Existence before essence." "Condemned to freedom."

Is There a Human Nature?

Where do we go from here? Since we are in the world of the Darwinian evolutionist, the obvious place to start is with Sartre's claim that there is no human nature. Together with his follow-up claim that there can be no human nature because there is no God to conceive it. In chapter 3, we were wrestling with people who were trying to convert Darwin's ideas into a value-laden religion or religion substitute. Let us go back to the spirit of the earlier chapters, where we showed how Darwin's theory refuted or put tension on various religious claims. It certainly puts tension on the claim that one can have a species with a nature only if it was so conceived by God. "And God said, Let the earth bring forth the living creature after his kind, cattle, and creeping thing, and beast of the earth after his kind: and it was so" (Genesis 1:24). What about evolution through natural selection? Was not the very title of the founding document the *Origin of Species*? To bring in the warthogs yet one more time—I never thought I would write a book where "warthog" would take top billing—evolution through natural selection made warthogs and evolution through natural selection made humans, and they are different—have different natures—because they live and thrive in different niches. Cause and effect. Biology tells all.

You must have been visiting Andromeda for the last decades if you are not aware that the term or concept *human nature* is, as they say, "much contested." One year, at the biennial meeting of the Philosophy of Science Association—a crowd of conventional academic thinkers if ever there was one—we were treated to an impassioned presidential address, where the very idea of human nature was declared anathema and cast into the hellfires of sociology or some other unfortunate social science (Hull 1986). Attitudes like this can be found not just in philosophy but across the spectrum of the humanities and social sciences and even into the physical sciences. Stephen Jay Gould did not deny the idea of human nature, but he sure didn't like the way in which it was tied exclusively to biology. Of the early movie of *Frankenstein*, where the monster is made innately evil, he wrote:

> Our struggle to formulate a humane and accurate idea of human nature focuses on proper positions between the false

and sterile poles of nature and nurture. Pure nativism—as in the Hollywood version of the monster's depravity—leads to a cruel and inaccurate theory of biological determinism, the source of so much misery and such evasive suppression of hope in millions belonging to unfavored races, sexes, or social classes. But pure nurturism can be just as cruel and just as wrong—as in the blame once heaped upon loving parents, in bygone days of rampant Freudianism, for failures in rearing as putative sources of mental illness or retardation or autism that we can now identify as genetically based, for all organs, including brains, may be subject to inborn illness. (Gould 1995, 99–100)

As Gould makes very clear, while the critical comments on ill-conceived notions of human nature may come from science, the motive is nigh always moral or social or political. It is felt that speaking of human nature, rather as one might speak of right-angled triangles, implies a uniformity that belittles or downgrades various groups of people—women, most obviously, but also LGBTQ and various racial groups. Not to forget the disabled. Think of the pressure toward getting prenatal testing for Down's syndrome.

Does Biology Really Have a Role?

I try not to be cowardly on these things. Indeed, I have a somewhat justified reputation for leading with my chin, having written on male-female differences, on race, and on sexual orientation (Ruse 2012). I wrote a whole book on homosexuality (Ruse 1988b). So, let me say some basic things and then, because I have a particular direction in which I want to go, move on. It is both obvious and good biology that evolution produces groups, the members of which share enough distinguishing features that we can consider them a kind or a type. The sorts of things that God was creating in the first week. There are good biological reasons for this, based on natural selection and the division of labor. Generally speaking, it is better to specialize than to be a jack of all trades. Birds stick to flying. Primates stick to the ground. Of course, there are exceptions— emus are more ground-bound than orangutans—but generally this holds, and biology leads you to expect exceptions. Don't make the exceptions more than they are. Human beings have two sexes— male and female. The fact that there are intersexes—way less than 1 percent—does not alter this fact. Humans are not triangles, and the fact that there are no exceptions to the right-angled triangle, square-of-the-hypotenuse theorem is irrelevant. (Actually, mathematics is more exception-prone than it usually lets on. As soon as you get into non-Euclidean geometry, many old favorites fall to the ground.)

Warthogs and humans are different, they have different natures. Warthogs are four-legged, they have snouts with horns, and they root around in the ground for food. Humans are bipedal, we are relatively hairless, we have comparatively very large brains so that, through the related large intelligence, we can go about our daily business. That said, within the groups or kinds or natures, there are going to be differences and variations. Variations that were put in place by natural selection and have adaptive virtues. Most humans are lactose-intolerant. They can digest milk as infants and then lose the ability. However, the evidence is that using milk is such a terrific adaptation that, with domestication of cattle and like animals, a genetic ability to utilize milk has been under strong selection for the past ten thousand years. The result is that dairy farmers like the Irish are nearly 100 percent lactose-tolerant, whereas those with no such practice are hardly tolerant at all. The Chinese are less than 5 percent tolerant. This is not at all to deny culture and how this can change and override biology. Dairy products became available because of the cultural move from hunter-gathering to agriculture. But remember that, in the end, culture exists because it is produced by our biology and is on balance biologically advantageous. Without culture, naked apes would not be able to live in the Arctic.

Lactose intolerance incidentally tells us that there are always exceptions, and culture cannot do better than tiptoe around them. After he got back from the *Beagle* voyage, Charles Darwin fell sick, and this plagued him for the rest of his life—headaches, stomach upsets, sleeplessness, and more. It is often hypothesized that this was all due to worry about his evolutionism, but that is unlikely. Darwin knew he was onto a winner there. More plausible is that Darwin's lifelong illness was due to lactose intolerance (Dixon and Radick 2009). Away at spas on a diet of thin gruel, and he was just fine. Back home and back to Emma's cream-based cuisine—we have her cookbook—and again he was sick as a dog. We would all agree, however, that Darwin was not only human but the paradigmatic instance of an Englishman—at one with the author on this vital matter. I, incidentally, can eat anything, although—unlike my younger daughter, who spent a summer in Vietnam—I have never eaten dog. If ever there were an absolute value, that is it for me. (Question posed on a vegetarianism-promoting billboard: "Why do we eat pigs but not dogs?" Answer by Emily Ruse: "Because they taste a lot better.")

Sexual differences and associated behaviors are the most obvious variation in humans. As it happens, there is still debate about the reasons for sexuality over asexuality—perhaps there is no one reason—but everyone agrees that it is something of adaptive value. The effort to find suitable mates—not to mention having males who do not themselves give birth—is outweighed by the virtues of having offspring with new and different combinations of genes. While not all of your kids will have the best adaptations, some will be in the running. What no evolutionary biologist is going to contest is that there will almost certainly be innate behavioral differences between the sexes. Mothers are mothers after all (Hrdy 1981; 1999). What no evolutionary biologist is going to contest is that how these differences play out might be very varied and culture will be involved. More middle-class, stay-at-home moms in Victorian England than in (the second) Elizabethan England.

Teasing out the biology from the culture is a difficult and ongoing task. A huge amount of theorizing in the past has been based more on prior cultural beliefs and expectations than on hard theory or evidence (Ruse 2012). Remember Darwin and his views on men as opposed to women. Much of what he said was simply false. Take intelligence. It is still much discussed if and what differences there might be, but from the point of view of raw IQ—a notion even more contested than human nature—few today would subscribe to the ordering of the Victorians. Of course, cultural change has been involved, but the biology was there all along. Changes in technology and attitudes and such things—women working in the world wars—have changed the expectations for women and opportunities. The cultural changes have taught us some home truths about innate qualities. When I started at university over fifty years ago, the ratio of men to women at the undergraduate level was about two to one. Every university administrator and professor now knows that the ratio is almost exactly reversed. Really, this should be no surprise. Raising babies is surely as complex as hunting warthogs.

Let me stress a little bit more the biological importance of our nature. One of the most striking things about humans is the extent to which males are involved in parental care. Of course, we all know of men who are selfish rotters when it comes to their partners and kids. Think of Abraham and Hagar and Ishmael. Note, however, that when Sarah put Abraham up to kicking out Hagar and their son, he was sad. "And the thing was very grievous in Abraham's sight because of his son" (Genesis 21:11). Men care about their children. Not just Ishmael, Abraham cared about Isaac. That is why the story of the potential sacrifice is so powerful. In the same mode, think of the story of the Prodigal Son and the love and concern of the father for both of his sons. What makes human behavior so striking is that this is all so uncommon among the mammals, including the higher apes. Male orangutans are just around for a good bout of sex, and then that is it. In the immortal words of Dean Martin: "Wham! Bam! Thank You Ma'am!" They are off, leaving the impregnated females to do the work on their own. There is no great secret or mystery about all of this. If it is in their biological interests to be around, males are going to be around. Not otherwise. Male birds help at the nest because, at the suitable season, the off-spring must be raised rapidly (Davies 1992). Up in trees, they are far more vulnerable than down on the ground. Or under-ground. If an impregnated mammal can just go down a burrow and stay there safe and sound, what need more has she of a male? Or so the males reason. We are all selfish genes. (Actually, we are not. More on that in a moment.)

Why then are human males so involved in child rearing? And why do we have physiological features to jolly this along? Think only of human continuous sexual receptivity. One suspects that male orangutan behavior would be very different if the females did not come into heat but one had to have ongoing sex to be certain of fertilization. One suspects that male human behavior would be very different if sexual intercourse were as pleasurable as having a root canal without anesthesia. The answer to our involvement is obvious. We have gone the route—no doubt part cause and part effect—of having offspring that need a huge amount of ongoing care in their early years. Fawns can get up on their feet within an hour of being born. Humans are taking early steps when they are one year old. Dogs and cats can already reason things out when they are ten weeks old. None of my children at the age of ten weeks showed much understanding that they are supposed to go outside to evacuate. (Actually, to be honest, neither do Cairn Terriers.) Fathers are not nice dads because of chance or because they were made that way by Jesus or because they heard the command of the Categorical Imperative—duty, duty, duty! They are that way be-cause their biology made them that way. Culture is also involved. My father made great sacrifices to send me away to boarding school when I was thirteen. I would never have done that to my kids. I liked teenagers! (Sometimes.) Divided by a generation gap, we were nevertheless both nice dads. Our biology made us so.

Our Social Nature

Now I get to the focus of my interest, how this all plays out with respect to meaning. I am going to say that Sartre is right, the meaning of life has to come from within. We are condemned to freedom. I am going to say also that Sartre is wrong, there is a human nature. It is rooted in our biology and then molded and informed by culture. This nature dictates what will be the meaning of life. Darwinian thinking is involved because it is so important in understanding human nature. You may say that this denies my claim to be an existentialist. I don't think existence precedes essence. I can live with that, although frankly, I don't think Sartre ever really thought that, either. If ever there was someone who was the quintessential Frenchman—habitué of the cafés of the Left Bank, cigarette in mouth and glass of wine in hand, non-stop earnest talking, overly brilliant in that irritating Cartesian way, never in bed except with someone to whom he was not married—it was he. But, said this author with some envy, less of him and more of me. To make good on my claims, I must concentrate on what I think is the most important aspect of human nature—one very widely shared by all peoples and all cultures. This is that we humans are a highly social species. "Human beings are not made to live alone. They are born into a family and in a family they grow, eventually entering society through their activity. From birth, therefore, they are immersed in traditions which give them not only a language and a cultural formation but also a range of truths in which they believe almost instinctively" (John Paul II 1998, 31).

We live all day and every day surrounded by our fellow species-mates—parents, children, students, co-workers, shop assistants, professional sports players, entertainers, actors on televisions or the stage, policemen and women, and so it goes on. Of course, there are exceptions, hermits and others, but they are exceptions. That is why solitary confinement is such an awful punishment. Aristotle knew the score. Writing in his *Nicomachean Ethics*:

> [Friendship] is either itself a virtue or connected with virtue; and next it is a thing most necessary for life, since no one would choose to live without friends though he should have all the other good things in the world: and, in fact, men who are rich or possessed of authority and influence are thought to have special need of friends: for where is the use of such prosperity if there be taken away the doing of kindnesses of which friends are the most usual and most commend-able objects? Or how can it be kept or preserved without friends? because the greater it is so much the more slippery and hazardous: in poverty moreover and all other adversities men think friends to be their only refuge. (Aristotle 1998, Chap. VIII)

Our social nature has been and still is one of humankind's strongest and most effective (biological) adaptations. Together we stand, divided we fall. It has been a cause-and-effect relationship. Being social worked, and so it intensified.

The results can be seen in at least two obvious ways. On the one hand, we have lost the features—formerly adaptive—that can mess up sociality. Most important, if you think about it, when it comes to fighting—Muhammad Ali notwithstanding—we are little more than sissies. The kind of chaps in the old Charles Atlas ads, where the strong guy kicked sand in our faces, and we stood helpless as he walked off with the pretty girl. That's just our fellow humans. We could not take on a lion or even your local gorilla—let alone a grizzly bear with a chip on its shoulder. On the other hand, we have gained the features—still adaptive—that can aid sociality. At the physiological level, once again, not coming into heat. At the best of times, it is tough enough trying to teach courses on symbolic logic. Imagine trying to do so if two of the students were in heat. At the psychological level, our big brains and large intelligences enable us to interact and to plan strategies and avoid pitfalls. They help to learn a lot of culture—starting with language—that makes us the social beings that we are. My dogs are pretty good at following instructions, but it would have been so much easier if I could just have talked to them and told them to avoid busy roads. And not to relieve themselves inside.

Family

I don't see anything particularly contentious or ideological here. Making note obviously of psychopaths and others, I am talking about all humans, in all places, at all times. Male, female, young, old, educated, ignorant, gay, straight, black, white, Christian, atheist. You may think me a little insincere in quoting John Paul II just above. In the same encyclical, he warned "against mistaken interpretations linked to evolutionism." Existentialism, too! By now, though, you surely realize that, whatever our differences, we are all on the same journey through the vale of soul-making. I have stressed again and again the extent to which evolutionary thinking is indebted to Christianity. Not to mention the Catholic adherence to natural law theory which stresses that in the Christian context we think and behave in natural ways—ways that a nonbeliever can and should recognize. As it happens, the pope was very favorable to evolution generally, and even Darwinism in particular (John Paul II 1997). So long as one gave him humans and their souls and so forth, which I suppose in the context of this discussion is somewhat like saying you are a little bit pregnant.

In a conciliatory spirit, however, I want to push the discussion a little further. How does our social nature manifest itself, how does it play out—especially in the light of our evolutionary nature and needs? Obviously, for any Darwinian evolutionist, reproduction must be at the top of the list. I don't mean we are all candidates for those expensive health farms that offer therapy for film producers with excess sexual energy. Equally, I certainly don't mean that people who don't reproduce are thereby inadequate. Without setting out deliberately to placate the pope, there are many plausible biological and cultural explanations to deal with this. I do mean that any major adaptation that does not speak directly to reproductive success is going to be short-lived. Obviously, here, though, there is no big issue. We humans have partners—reasonably monogamously (perhaps less monogamously than when we were poorer and died much earlier) or stably polygamously—we have children for whom we care, we have families, and sometimes these are quite extended, grandparents or children, siblings and their families, aunts and uncles, cousins, and more. Often the non-reproducers—maiden aunts, for instance—play major roles in the family's well-being.

Two points. First, we don't do all of this because of the stern call of the Categorical Imperative. Duty, duty, duty. I won't say it never occurs. As anyone who has had kids can well testify, things have a nasty way of going wrong. Bad grades, unhappy love affairs, job prospects growing dimmer and dimmer. Sometimes, having those long, supportive, middle-of-the-night phone calls is duty, duty, duty. Hume knew this:

> In like manner we always consider the natural and usual force of the passions, when we determine concerning vice and virtue; and if the passions depart very much from the common measures on either side, they are always disapproved as vicious. A man naturally loves his children better than his nephews, his nephews better than his cousins, his cousins better than strangers, where everything else is equal. Hence arise our common measures of duty, in preferring the one to the other. Our sense of duty always follows the common and natural course of our passions. (1739–40, 483–484)

Truly, though, Hume is right. What motivates my wife and me is that we love our children dearly and are hugely proud. Grades improve, partners turn out to be human beings after all, and suddenly there is work on Monday morning and pay on Friday afternoon. Emotions and feelings are the big thing. We love our kids and our spouses and our parents—even our siblings sometimes! "He ain't heavy, Father . . . he's m' brother." Second, there are good biological models to explain all of this. Darwin didn't have genetics, so he couldn't see the big picture, but he grasped that inasmuch as relatives reproduce, so also do you—sharer of the same units of heredity, genes—by proxy. Not as well directly as you do yourself, except for identical twins, but better than nothing. Sometimes, if you can get enough relatives to reproduce, you are better employed doing it by proxy than doing it yourself. The Hymenoptera (ants, bees, and wasps) are even further in this than humans. Because of a funny mating system—females have both parents, whereas males have only mothers—females are more closely related to sisters than to offspring. The sterile workers reproduce most efficiently by caring for the queen and raising her offspring. In the 1960s, William Hamilton (1964) formalized a lot of this, and John Maynard Smith (1964) called it "kin selection." There is nothing strange or other than ultra-Darwinian processes going on here. Maiden aunts do not make Darwinians despair!

Morality

Moving beyond family, Aristotle is right. We have social relations with friends and beyond. I don't know the chap in the supermarket, but he directs me to the ethnic section where I can find brown sauce, English-style, not to mention Marmite, a vile-smelling, grotesquely salty yeast spread that goes on toast in the morning and that is truly addictive for those of us raised on it. I, in turn, take up my wares and go to check out and pay. Ecologically sensitive as I am—not to mention being the spouse of a ferocious wife of whom I am terrified—I put it in the cloth bag I brought from home for the purpose. We work and interact socially. In this global village, it spreads out beyond. These are usually not blood relatives. One powerful force behind this is what is known as "reciprocal altruism"—you scratch my back, and I will scratch yours (Trivers 1971). Darwin spotted this and talked about it in *The Descent of Man*. "In the first place, as the reasoning powers and foresight of the members became improved, each man would soon learn that if he aided his fellow-men, he would commonly receive aid in return. From this low motive he might acquire the habit of aiding his fellows; and the habit of performing benevolent actions certainly strengthens the feeling of sympathy which gives the first impulse to benevolent actions" (1871, 1, 163–164). As Darwin saw full well, you are not going to interact all the time having first calculated the odds of return. Should I help this child who has wandered onto a busy road? Will his parents give me something? Must I do this in order that others will do it for me and my children? We need a quick and dirty solution for getting instant action. Sometimes it won't pay off. I get killed, too, and people think me silly not to have flagged the oncoming driver instead. But overall it does pay off. If not now, then later. And if not later, well, it was still good insurance. It is here that morality comes in. Evolution had given us not just sentiments of love and friendship but also that sense of duty. Hume saw it. Kant obsessed about it. There are some things we should do and some things we should not do. Of course, it is bound up with culture. You ought to open boarding schools for the middle classes. You ought not open boarding schools for the middle classes. Ultimately, it is sentiment with the added ingredient of obligation (Ruse 1986; Ruse and Richards 2017).

At this point many, Christians but others also (like Plato and G. E. Moore), question whether you can get morality without God or some external source, like that which confers eternal truths on mathematics. How can "Oppose the Nazis," which led Sophie Scholl to her death, be just evolutionary sentiment? Two points. First, no one now is trying to justify morality. Just say that it exists. We are not up to the tricks of the last chapter. I am a good Humean. To use a sports metaphor, rather than (as in chapter 3) smashing through the is/ought barrier, I want to do an end run around it. I am not justifying morality on the basis of evolution. My own opinion is that morality has no justification and can have no such justification. Given the non-directedness of Darwinian evolution, I believe we could believe something completely different if things had been otherwise. Darwin puts the matter graphically: "If, for instance, to take an extreme case, men were reared under precisely the same conditions as hive-bees, there can hardly be a doubt that our unmarried females would, like the worker-bees, think it a sacred duty to kill their brothers, and mothers would strive to kill their fertile daughters; and no one would think of interfering." Continuing: "The one course ought to have been followed, and the other ought not; the one would have been right and the other wrong" (Darwin 1871, 1, 73).

Second, whether or not morality is subjective, part of the personal experience of morality is that it is not just a matter of choice but is objective. As far as we are concerned, this is genuine morality and not just enlightened self-interest. It would be gro-tesque to say that Sophie Scholl acted out of self-interest. Even if she hoped her actions would get her into heaven, nothing detracts from her saintly behavior. I said in the first chapter that natural selection has a way of getting up people's noses, and I suspect here is a major reason why. For all that research has shown that much of this thinking is a projection back from the twentieth century, there is fear that we are on the road to the supposed ruthless Social Darwinism of nineteenth-century industrialists (Ruse 2017c). It is true that modern evolutionary theory is, in the tradition of Darwin himself, wary of selective effects that do not rebound to the individual but only to the greater group. Critics worry whether this "selfish gene" approach precludes genuine morality—even the experience and conviction of genuine morality. This fear of "bio-logical reductionism" is felt by the extreme left, for instance, the Marxist biologist Richard Lewontin (1991), and the extreme right. The conservative philosopher Roger Scruton writes: "It makes cynicism respectable and degeneracy chic. It abolishes our kind—and with it our kindness" (2017, 49). For this reason, critics push for an extended non-Darwinian reading of selection, where the unit is the group rather than the individual. Many would prefer to get away from selection entirely and look for another mechanism, perhaps an updated Lamarckism. Something warmer, more holistic, or whatever (Reiss and Ruse forthcoming).

This is a needless worry. On the one hand, it is not because they are cynical egoists that today's evolutionists follow Darwin in focusing on the individual rather than the group. It is simply the technical issue that they fear the problem of cheating. If selection does favor the group over the individual, then cheating pays—you get group help but give nothing—and morality soon collapses as everyone becomes a cheater. Holism and so forth is just fine. Actually, it isn't always (Harrington 1996). The greatest holists of the twentieth century were the National Socialists: *Ein Volk, Ein Reich, Ein Führer*. Nice or nasty, holism is more a philosophical aspiration than serious science. On the other hand, selfish genes are a metaphor (Dawkins 1976; Ruse 1979b). There is no reason to think individual actors are selfish. As pointed out above, one can readily think of reasons why we would function more efficiently if we really believed in the call of disinterested altruism than if we were scheming always to get the most for ourselves. Don't start calculating your benefits. Save the kid! At the level of human awareness and action, the whole point is that we are anything but selfish. Most of the time we are not facing Sophie Scholl choices. We are just trying to get on with our daily lives. Care about your students. Don't cheat on your taxes. Don't mark up library books. What or who is to stop me? The still, small voice of conscience. "And it was so, when Elijah heard it, that he wrapped his face in his mantle, and went out, and stood in the entering in of the cave. And, behold, there came a voice unto him, and said, What doest thou here, Elijah?" (1 Kings 19:13). I bet he was careful with his library books after that.

The Life of the Mind

I suspect that there will still be a strong sense of unease, that in some way I am missing the elephant in the room. Consciousness, self-awareness, introduces a whole new dimension, and at this level biology is irrelevant, perhaps misleading. Scruton, after a sympathetic introduction to the biology of human nature, writes bluntly: "I want to take seriously the suggestion that we must be understood through another order of explanation than that offered by genetics and that we belong to a kind that is not defined by the biological organization of its members. The 'selfish gene' theory may be a good account of the origin of the human being: but what a thing is and how it came to be are two different questions, and the answer to the second may not be an answer to the first" (2017, 19).

At one level, I agree entirely. David Copperfield's reactions after meeting Dora are very different from what drives a drone to fly ever higher to mate with the queen. It is just that—quite apart from issues about the nature of the mind-body relationship, of which a little more later—I don't see the biological and the cultural as rivals, as Scruton does, but much more as complements, fellow laborers in the vineyard. David and the drone are both in the reproduction business. It is just that they go about it in different ways. Very different ways. I have elsewhere suggested that adding consciousness to the mix is a little like adding the complex number system to algebra. You have been told all along that minus numbers cannot have square roots, and now here it is! At least it has the decency to call itself an "imaginary number." Soon, however, you learn that i is really not so very different—it can be added and multiplied—and is so very powerful. For a Darwinian, consciousness has much the same role. Can't be done, but it is, and is subject to many of the same rules—two reasons are better than one, that sort of thing—and so very much more powerful. This said, if we are not serving the end of biology in some ways—often ways more restrictive than we imagine—it isn't going to work (Ruse 2017b).

Picking up the thread of the argument, showing even more how little in important respects I differ from someone like Scruton, agree that, after family, society in general and its moral underpinning are part of the adaptive package deal that characterizes human nature. There is a third aspect. Here, I really am not sure how much it is directly biologically adaptive and how much something extra brought on by culture, what Gould called an "exaptation" (Gould and Vrba 1982). I am referring to the whole world of the mind—particularly the creative mind—science, literature, music, art, politics, and so much more. The truly rich side of the culture in which we live. How much of this is directly biological? Obviously, some. Technology is how we survive and reproduce in most parts of the world—actually, in all parts of the world. Even the most "primitive" of societies have not only language but remarkably sophisticated ways of surviving. Darwin thought music was bound up with sexual selection. Thomas Hardy, in *Tess of the D'Urbervilles*, makes much of this. Angel plays his harp:

> Tess had heard those notes in the attic above her head. Dim, flattened, constrained by their confinement, they had never appealed to her as now, when they wandered in the still air with a stark quality like that of nudity. To speak absolutely, both instrument and execution were poor; but the relative is all, and as she listened Tess, like a fascinated bird, could not leave the spot. Far from leaving she drew up towards the performer, keeping behind the hedge that he might not guess her presence. (Hardy 1892, 146–147)

"Stark quality like that of nudity"! Little wonder that again and again Hardy had to bowdlerize his tales to get them published.

I want to push this point a little bit. The life of the mind has direct adaptive value in helping us to explore and manipulate our environment. The point I make here is that this life of the mind is intensely social, whether as a direct adaptation or as a spin-off. Think of the science we have introduced into this essay. If anything is a group activity, it is science. It is true that Darwin worked in secret on his theory, but it was all made of elements he got from others, and as soon as he felt able he was passing it on to others. It is the same in other fields. It's all about relationships, both in content and in form. Think Shakespeare. Romeo and Juliet and their doomed love, set against the terrible relationship between their parent clans. Think music. I love opera and none more than the Mozart/da Ponte opera *Cosi Fan Tutte*. It tells the tale of a cynical old man—a philosopher!—who teaches two silly young pairs of lovers a lesson in the shallow nature of so much of our emotions and relationships. The men place a bet on their girlfriends' fidelity and, before long, find themselves disguised and each courting the girl of the other. Egged on by Despina the maid—who is happy to mess up their love lives for a coin or two from the philosopher, Don Alfonso—you can guess what happens. Silly? Profound? Immoral? Cynical? It has been called all these things. As with Shakespeare, it is deeply social, both in content and in the fact that it is done for the entertainment of others. Art, too. After years of condescension, an exhibition at the Art Institute in Chicago won me over completely to the conviction that Roy Lichtenstein, with his pop art, was one of the most important American artists of the twentieth century. It is all about relationships. "I love you, too, Jeff, but." We think all the time as social beings. And never more than in what at first seems so solitary, the life of the mind. If you are reading this, you make my point!

Killer Apes?

With respect to meaning, you can see easily where I am going with all of this. I am going to tie meaning right in with our Darwin-evolved human nature. First, however, let me speak to an obvious objection (Ruse 2018b). Stop it with all the sociality. Humans are not only naked apes, they are killer apes. We are born with bloodlust, and we live and die with it. Two world wars, Korea, Vietnam, and now we have an endless war in Afghanistan. Darwinians know the score and know that violence is part of our nature. Konrad Lorenz, a father of ethology, was very clear on this. In animals, there are all sorts of mechanisms that stop violence from escalating within the species. There is conflict, true. At some point, however, something physiological kicks in and stops the progress to killing and death. Often, the loser or the weaker will show submissive behavior. This does not lead to instant death but to a diminution of aggression and violence. "This is certainly the case in the dog, in which I have repeatedly seen that when the loser in a fight suddenly adopted the submissive attitude, and presented his unprotected neck, the winner performed the movement of shaking to death, in the air, close to the neck of the morally vanquished dog, but with closed mouth, that is, without biting" (Lorenz 1966, 133).

Humans, alas, have gone badly wrong:

> In human evolution, no inhibitory mechanisms preventing
> sudden manslaughter were necessary, because quick killing
> was impossible anyhow; the potential victim had plenty of
> opportunity to elicit the pity of the aggressor by submissive
> gestures and appeasing attitudes. No selection pressure arose
> in the prehistory of mankind to breed inhibitory mechanisms
> preventing the killing of conspecifics until, all of a sudden,
> the invention of artificial weapons upset the equilibrium of
> killing potential and social inhibitions. When it did, man's
> position was very nearly that of a dove which, by some un-
> natural trick of nature, has suddenly acquired the beak of a
> raven. (241)

Killer apes indeed!

This was written over fifty years ago and there has been a sea
change in our thinking on this subject. For a start, it turns out
that a huge amount of this stuff about our innate aggressive nature comes not from nature but from the Bible, or at least the
Bible as interpreted by Saint Augustine. Hugely influential was
Raymond Dart (1953), the South African discoverer of Taung
Baby, *Australopithecus afarensis*. He was backed by his American
popularizer, Robert Ardrey (1961). Their sources were Augustine
on original sin and the hypothesis about how in order to wipe
this away there was need for Jesus to die on the cross. A blood
sacrifice to assuage God. This dark scenario, quasi-naturalized, fit
happily with its time in history, after the Second World War and
Korea, in the middle of the Cold War, and with Vietnam on the

horizon. No one had any empirical evidence, so let us go with the long-believed.

The evidence has started to come in (Tuttle 2014). It is now thought by both evolutionary biologists and cultural anthropologists that our ancestors were far more peaceable (Fry 2007). That is why we lost so many of our natural weapons. It was only with the coming of agriculture, about ten thousand years ago, that the need and opportunities for conflict arose and war started to become the norm. Even here, without exaggerating or minimizing the conflicts of the last century, statistically the deaths from conflict and violent behavior have declined significantly over the years. Even in America, with its murderous gun policies. Harvard linguist and evolutionary psychologist Steven Pinker writes: "Believe it or not—and I know that most people today do not—violence has declined over long stretches of time, and today we may be living in the most peaceable era in our species' existence." This has had significant consequences: "Daily existence is very different if you always have to worry about being abducted, raped, or killed, and it's hard to develop sophisticated arts, learning, or commerce if the institutions that support them are looted and burned as quickly as they are built" (2011, xxi).

What brought all of this on? We are a mixed bag. Sometimes violent, sometimes not. This is the legacy of our evolution. The question is how the non-violent side has gained the upper hand. "The way to explain the decline in violence is to identify the changes in our cultural and material milieu that have given our peaceable motives the upper hand" (Pinker 2011, xxiii). Pinker identifies several factors. Showing how none of this is yet totally settled, he is inclined to think that the move to agriculture brought a lessening of violence rather than a rise. Pinker is nothing if not controversial, and I have myself had a go at jabbing at him. Perhaps philosophers still have the taint of the killer ape! More seriously, the fact is that we are violent at times, but this does not deny our essentially social nature. In wartime, the buddy system is all-important.

Three Things of Great Worth

What then do we say about meaning in this Darwinian world? Simply, that the meaningful life—as Aristotle argued strongly— is to live our human nature properly, to the full. That is what nature designed us to do, and the only way we can live happily, find true contentment, is to do what we have been told. We are free to choose, but don't ignore our evolved human nature. First, family. Again, let me stress I am not now arguing against people who do not have children. There are many reasons why this may be so, starting with infertility and going on to circumstances. I remember, when I started out on a career in philosophy, how much easier it was for men than women to marry and have children. This, I am glad to say, has much changed. The fact is that family life for humans is deeply meaningful. Loving and creative. I am sure many of my readers can resonate with that. After a very unhappy first marriage, I met and married Lizzie, and we have had and raised children. As our children show only too well, it is not always easy, but it is always worthwhile and important.

Second, the moral life in society, close and distant. I say to my children and my students: the only truly happy person is the person giving to others. There is no magic to this. It is human nature. Think of Plato's *Republic*. Thrasymachus, whose philosophy is "might is right," is refuted. Most of all by the alternative world that Plato sketches out. Admittedly, the first call on the philosopher kings is the good of the whole state, but they would never know true happiness—participate in the Form of the Good—if they did not heed this call. It is still true today. Ask yourself who are the happy people among your friends and acquaintances, and who are much less so. Then ask yourself who are the givers. The person who stays late to help a student. The person who turns out on a Saturday to join the park clean-up crew. The person who always buys the first round. Ask yourself about the others. Have to get home now. Sorry, Saturdays we do the shopping. I'll buy the first round next time. The conclusion follows readily. The groups are the same. Givers are happy, non-givers less so. No magic about this. It is human nature.

Third, the life of the mind. You may find this book boring, but I don't! I just love exploring ideas through time and space. Being creative yourself and enjoying the creative fruits of others. I cannot imagine a life without Charles Dickens. Time for another extract from *David Copperfield*, one that I read first when I was about ten. Much younger than in the last extract, David has been sent off to school and has a dinner ordered for him at an inn. Overwhelmed by the food and the strange experience, David lets a too-friendly waiter eat it all for him:

"There was a gentleman here, yesterday," he said—"a stout gentleman, by the name of Topsawyer—perhaps you know him?"

"No," I said, "I don't think—"

"In breeches and gaiters, broad-brimmed hat, grey coat, speckled choker," said the waiter.

"No," I said bashfully, "I haven't the pleasure—"

"He came in here," said the waiter, looking at the light through the tumbler, "ordered a glass of this ale—WOULD order it—I told him not—drank it, and fell dead. It was too old for him. It oughtn't to be drawn; that's the fact."

I was very much shocked to hear of this melancholy accident, and said I thought I had better have some water.

"Why you see," said the waiter, still looking at the light through the tumbler, with one of his eyes shut up, "our people don't like things being ordered and left. It offends 'em. But I'll drink it, if you like. I'm used to it, and use is everything. I don't think it'll hurt me, if I throw my head back, and take it off quick. Shall I?"

I replied that he would much oblige me by drinking it, if he thought he could do it safely, but by no means otherwise. When he did throw his head back, and take it off quick, I had a horrible fear, I confess, of seeing him meet the fate of the lamented Mr. Topsawyer, and fall lifeless on the carpet. But it didn't hurt him. On the contrary, I thought he seemed the fresher for it.

Sorry for the length of the quotation. Actually, no apologies at all. Once I had started reading, I never stopped and am going on still. Music, too. After a great opera performance—a production I saw recently in Santa Fe of *Lucia di Lammermoor*—my feeling was: God, I don't care if you exist or not. We won! In philosophy, too. At the age of eighteen, at school, instead of scripture—mine was a private Christian school but sharing the conviction that all education should be directed toward the needs of running the Empire—we read the *Republic*. If that isn't a book about human beings and their social relationships, I don't know what is. My life was changed. No empires, thank goodness.

What's the Alternative?

There you have my formula for the meaningful life in the Darwinian world—family, friends and others, works of the creative imagination—and I mean science here as much as the arts. Let me just turn it the other way for a moment. What if you still have yearning for Meaning in a big way, as offered by religion? As one who was brought up intensely religious, I would be a liar if I denied that there is always that lingering trace of sadness at hope extinguished. But ask yourself for a moment how the promises of religion would play out, given what we now know of our Darwinian-produced human nature. Buddhism would certainly keep us busy and occupied through the incarnations, even though our nature might not be human very often. The endpoint, though, seems to have the same problems as Christianity. Boring! What the philosopher Bernard Williams (1973) called the "tedium of immortality." The thought of eternity sitting on a cloud wearing a bedsheet playing a harp is frightening. After a few billion years, fully conscious root-canal procedures would have nothing on it.

You might say that this is unfair. Heaven is going to be an ethereal thing, ecstasy with the Creator. Well, for a start, this isn't exactly what Saint Paul says. He clearly believes in some kind of bodily resurrection. "It is sown a natural body; it is raised a spiritual body. There is a natural body, and there is a spiritual body" (1 Corinthians 15:44). For a second, it is hard to know quite what an eternity of ecstasy might mean. Is it like an ongoing sexual climax? I think this idea appealed to me more as a teenager than it does now. Without wanting to sound a complete old fuddy-duddy, there are times when I find more attractive the thought of a nice chat over a cup of tea. Are we going to have to change jobs on a regular basis to avoid the tedium of eternity? "Right, Ruse, for the next millennium you are going to be a lavatory attendant." (Assuming spiritual bodies have need of lavatory attendants.) Are we going to change spouses? Are we going to have spouses? Saint Matthew assures us: "For in the resurrection they neither marry, nor are given in marriage, but are as the angels of God in heaven" (Matthew 22:30). No, thanks very much.

You may think I am joking, and I am in a way. I am also deadly serious. Don't misconstrue my irreverence the wrong way, even thinking that it is irreverence. Given human nature—so entirely temporal—I simply don't know what eternity would be like. Or could be like. An eternity without Lizzie is too awful to contemplate. I don't mean to be maudlin or sentimental, but for nearly forty years now, our lives have been together, loving. Two minds, yes, but somehow two minds that are one, sharing the joys and the troubles. We quarrel—I am not exactly truthful about the money I spend on books—but even in the quarrels we are together. In the same mode, I feel much the same about my little Cairn Terrier, Scruffy McGruff. We do have serious discussions about armadillos and their vulnerability. And what about birds and trees and much more? Don't misunderstand me. I don't think I have an I-thou relationship with my dog, although one wag once described it as a "bow-wow" relationship. That about sums it up. Ed Wilson (1984) is good on this. We evolved in symbiotic relationship with the rest of the living world. We need it spiritually as much as we need it physically. It is part of what it is to be a human. I am glad to note that, for all the problems I have with the Christian God, He is sensitive to at least part of this problem. "Are not two sparrows sold for a farthing? and one of them shall not fall on the ground without your Father" (Matthew 10:29).

At this point, I suspect many Christians will nod in agreement and say that they, too, don't know what eternity will be like. To say again what has been said before, ultimately, it is all cloaked in mystery. Thus, John Paul II: "Revelation remains charged with mystery. It is true that Jesus, with his entire life, revealed the countenance of the Father, for he came to teach the secret things of God. But our vision of the face of God is always fragmentary and impaired by the limits of our understanding" (1998, 13). Thus, the Apostle Paul: "For now we see through a glass, darkly; but then face to face: now I know in part; but then shall I know even as also I am known" (1 Corinthians 13:12). In a major sense I agree. I think in the end we simply don't know. At the risk of becoming repetitive—no more warthogs!—I quote Richard Dawkins one more time: "Modern physics teaches us that there is more to truth than meets the eye; or than meets the all too limited human mind, evolved as it was to cope with medium-sized objects moving at medium speeds through medium distances in Africa" (2003, 19).

All the more remarkable given the source from which it comes, that really is a profound statement. Be modest about what you know or can know. You think you can find out everything, but that ain't necessarily so. Complementarity—electrons as particles, electrons as waves—is not solved by the Heisenberg Uncertainty Principle. It is just ruled off limits for scientific questions. Why is there something rather than nothing? This is a genuine question, but with no answer or conceivable answer. Even worse is the problem of consciousness. We know a lot about the brain and about body-mind interactions, but consciousness itself—sentience—I think therefore I am—we haven't solved it. We haven't even scratched at solving it. As I said before, although we have not yet solved the natural creation of life, I know what a solution would look like. RNA worlds, deep sea vents, and so forth. In the case of consciousness, I don't even know what a solution would look like.

Panpsychism and Its Discontents

I am not now being an intellectual Eeyore. More important, I am not now wimping out on taking a path that possibly takes me back to the plausibility of Darwinism as religion. Philosophically, I think we can make some progress on the mind-body problem. Cartesian dualism, *res extensa* and *res cogitans*, is attractive. It cannot be true. Mind and body are not independent. Materialism, à la Dennett? I have already been rude about that. It is just silly to say that we don't think or that when we think it is nothing but molecules in motion. Popular in many circles is some version of emergentism (Humphreys 2016). You put the molecules together in a certain way, and—hey, presto—you have thought. Popular but unconvincing. It simply reduces the whole thing to a miracle. Pythagoras's theorem emerges from putting together three straight lines in a certain way. You start with geometrical entities, and you end with geometrical entities. The whole point about the consciousness issue is that you have something completely different.

Increasingly high-profile today is some kind of panpsychism, thinking in the mode of Spinoza that ultimately mind and body are one substance (Brüntrup and Jaskolla 2016). You are not now claiming, with Prince Charles, that plants can think. You are saying that consciousness in some form is part of all existence—as one might say it is in some sense of us when we are in dreamless sleep. There is some fascinating suggestive evidence, most striking being quantum entanglement. Something happening on one side of the universe has a mirror happening on the other side of the universe, simultaneously. It cannot be causal in the usual sense of the term, but it is a transfer of information. And what is consciousness but information in some way?

I am attracted to panpsychism because it takes evolution seriously. The nineteenth-century mathematician-philosopher William Kingdom Clifford wrote:

[W]e cannot suppose that so enormous a jump from one creature to another should have occurred at any point in the process of evolution as the introduction of a fact entirely different and absolutely separate from the physical fact. It is impossible for anybody to point out the particular place in the line of descent where that event can be supposed to have taken place. The only thing that we can come to, if we accept the doctrine of evolution at all, is that even in the very lowest organism, even in the Amoeba which swims about in our own blood, there is something or other, inconceivably simple to us, which is of the same nature with our own consciousness, although not of the same complexity. (Clifford 1879, 2, 38–39)

Even if panpsychism appeals philosophically—and I like it if only because it fits in with what I said about sentience being like something that, for all its strangeness, is part of the system—I am not sure that this takes you much further toward getting a good scientific answer to the nature of consciousness. I am certainly here not offering a measured defense of panpsychism. Apart from anything else, I am not solving the "combination problem"—how do you get all those mind-infused separate molecules that make up Michael Ruse into a single thinking, sentient being (Chalmers 2014)? I am a little less worried by that problem than perhaps I should be. Although I have been a bit rude about holism, I am not against it in principle—or in practice, for that matter. I just don't think it is a priori preferable. It depends. Sometimes go small, reductionism. Sometimes go big, holism. Quantum entanglement is suggestive of holism, and perhaps there is a clue there about things coming together (Ismael and Schaffer 2016). We are no longer going from chalk to cheese, from *res extensa* to *res cogitans*—the miracle of emergence. We are going from one level of mind to another level of mind—straight Euclidean lines to Euclidean triangles. No miracle there. No miracle perhaps, but much more work to be done. Perhaps it will help if we start taking seriously what the physicists tell us, that today *res extensa* is no longer thought of as chunks of matter but in some vital sense involves waves and so forth. Waves do combine. Try using your foot to slosh the water up and down in the bathtub. As the water comes surging out and dripping down, see how quickly your mother will have the broomstick pounding on the ceiling from the floor below.

Pertinent here, I don't see panpsychism in itself getting you back to objective comparative values, a gussied-up form of Darwinism as religion. You are still where we were in chapter 3. Agree that value is associated with consciousness. Cartesian *res extensa* can never have value in its own right. You must impute value to it—God or humans. *Res cogitans*. Consciousness, even a lesser kind than human, does give value. If you are going to let all of creation have the rudiments of consciousness, then value surely follows. Those mountains around Stellenbosch really do have value in their own right. Fair enough, but the same problems arise as before. Unlike Descartes, I do think Scruffy McGruff has a level of consciousness. I think also he has value in his own right. I value him very much, it is true, but much of this value is value I find. I just don't ascribe it. Being a subjectivist, in a sense I think we do ascribe all value. Within my system I don't necessarily. I could play baseball rather than cricket; having opted for cricket, I must bowl and not throw at the batter. As captain, I still have the choice about whether to use a pace or a spin bowler. Within my system, *Cosi Fan Tutte* has objective value. Marmite—just ask my wife—does not. I do not even dare try it out on graduate students. There was once a TV ad of children fleeing an approaching jar of the stuff, so terrifying that the authorities made it be taken off the air. The manufacturers of Marmite claim one's taste is an objective matter. Like lactose intolerance, some people don't have the right genes. Jars of Marmite carry a picture of the DNA model, the double helix.

Ascription or not, we still have the problem raised in chapter 3. Why me over Scruffy? Why Scruffy over the mountains? Don't say because I have more of the mental stuff than Scruffy and he has more than the mountains. Even if that makes sense and is true, Darwinism is no help here. It is the Jack Sepkoski problem all over again. I'd bet on the mountain before I would bet on me. Really, in the end it would all be a question of the ordering by that Being who chooses humans over dogs, dogs over mountains. "Fear ye not therefore, ye are of more value than many sparrows" (Matthew 10:31). Which leads to the nasty suspicion that that Being might be me. I first put in what I then pretend to discover. All books in my world have value, but some are to be valued more than others. Why? Because that is my decision. It is I and I alone who puts my first edition of the *Origin* above all others. (Just joking, I am afraid. I do have a first edition of *Evolution: The Fossils Say No!* by Duane T. Gish, PhD, signed, "To Michael Ruse, with warm personal regards." Hope springs eternal in this bibliophile that Gish was right and Darwin was wrong.)

Leave things at that—although I do confess to a nagging worry about whether, over Scruffy, I would choose my late headmaster, who contrary to all appearances I am assured was human. In the end, as I said in chapter 2, in an important sense I end as what Colin McGinn (2000) has called a "new mysterian." I don't know the answer to consciousness, and I rather suspect that our limited, Darwinian-evolved brains are not going to solve it for us. It is a problem that will be forever beyond our ken. Whether or not this is true, it doesn't seem that objective meaning is to be found that way. Even if you want to say that everything has value, there is no troublesome violation of Hume's Law. No getting the absolute comparative values that the religious—spiritual or secular—want out of Darwin's theory. Four legs good, two legs better. You don't learn that from science.

Agnosticism

What does this have to do with Meaning? In the end, I can give you a good Darwinian account of Meaning in terms of our evolved human nature. This is not a weak substitute. This is the real thing. I have worked hard in my life to do what I do—raise five children, teach for over fifty years, write more books than it is decent to count. I have found it immensely satisfying. I see no reason to expect anything beyond this. From an eternity of oblivion. To an eternity of oblivion. Everlasting dreamless sleep, without the need to get up in the middle of it to go to the bathroom. Absurd if you will, although I would not call it this. In the end, though, I am an agnostic. I just don't know whether life has any—time for those capitals—Ultimate Meaning. After all, for all that, in the *Phaedo*, Socrates rather implies otherwise, even dreamless sleep doesn't at once translate into non-being.

Does my ignorance put me in the same camp as someone like John Hick? Appreciating the ineffable, the unknowable? I should tell you that Hick and I attended the same Quaker school, although separated by years, so perhaps there is a reason for a sense of sameness. That shared sense of mystery. We are not really on the same page, though. As a teenager, he had a road-to-Damascus experience that took him to Jesus. As a teenager, I was already starting my move away from Jesus. Theologically, we really do differ. Hick thinks his life force exists and is a force for the good. I am far from certain that that exists, and if it does exist, I think it could easily be coldly indifferent—one of the "purblind Doomsters."

No, my position is not a cop-out, trying to sneak in God at the end. Cross my heart and hope to die—I will certainly get that wish—I just don't know. I am with J. B. S. Haldane on these things: "My own suspicion is that the Universe is not only queerer than we suppose, but queerer than we can suppose" (1927, 286). There may be something more. There may not. Don't spend your life agonizing about this or letting people manipulate you with false promises. Think for yourself, as my Quaker mentors insisted. Life here and now can be fun and rewarding, deeply meaningful. Remember, Hume didn't just play backgammon—he dined, he conversed, he was "merry with my friends." Like I said: a nice cup of tea, or perhaps a single malt, and a chat. With my beloved graduate students and Scruffy joining in the conversation! Live for the real present, not the hoped-for future. Leave it at that.

> For that life is dear,
> The lust after life
> Clings to it fast.
> For the sake of life,
> For that life is fair,
> The lover of life
> Flings it broadcast.

The lover of life knows his labour divine,
And therein is at peace.
The lust after life craves a touch and a sign
That the life shall increase.

The lust after life in the chills of its lust
Claims a passport of death.
The lover of life sees the flame in our dust
And a gift in our breath.

(Meredith 1870, 182–183)

BIBLIOGRAPHY

Anscombe, G. E. M. [1957] 1981. Mr. Truman's degree. *Ethics, Religion and Politics: Collected Philosophical Papers*, Vol. 3. Editor G. E. M. Anscombe, 62–71. Oxford: Blackwell.

Anselm. [1077–1078] 1903. *Anselm: Proslogium, Monologium, An Appendix on Behalf of the Fool by Gaunilon; and Cur Deus Homo*. Translator S. N. Deane. Chicago: Open Court.

Aquinas, T. [1265–1274] 1952. *Summa Theologica I*. London: Burns, Oates and Washbourne.

———. [1259–1265] 1975. *Summa contra Gentiles*. Translator V. J. Bourke. Notre Dame: University of Notre Dame Press.

Ardrey, R. 1961. *African Genesis: A Personal Investigation into the Animal Origins and Nature of Man*. New York: Atheneum.

Aristotle. 1984. *Physics. The Complete Works of Aristotle*. Editor J. Barnes, I: 315–446. Princeton, NJ: Princeton University Press.

———. 1998. *Nicomachean Ethics*. Editor J. A. Smith. Mineola, N.Y.: Dover.

Augustine. [397–400] 1998. *Confessions*. Translator H. Chadwick. Oxford: Oxford University Press.

Ayala, F. J. 2009. Molecular evolution. *Evolution: The First Four Billion Years*. Editors M. Ruse and J. Travis, 132–151. Cambridge, Mass.: Harvard University Press.

Bada, J. L., and A. Lazcana. 2009. The origin of life. *Evolution: The First Four Billion Years*. Editors M. Ruse and J. Travis, 49–79. Cambridge, Mass.: Harvard University Press.

Barbour, I. 1990. *Religion in an Age of Science*. New York: Harper and Row.

Barrett, P. H., P. J. Gautrey, S. Herbert, D. Kohn, and S. Smith, editors. 1987. *Charles Darwin's Notebooks, 1836–1844*. Ithaca, N.Y.: Cornell University Press.

Barrow, J. D., and F. J. Tipler. 1986. *The Anthropic Cosmological Principle*. Oxford: Clarendon Press.

Baynes, T. S. 1873. Darwin on expression. *Edinburgh Review* 137: 492–508.

Benatar, D. 2017. *The Human Predicament: A Candid Guide to Life's Biggest Questions*. Oxford: Oxford University Press.

Benn, A. W. 1906. *The History of English Rationalism in the Nineteenth Century*. London: Longmans, Green.

Bergson, H. 1911. *Creative Evolution*. New York: Holt.

Bowler, P. J. 1988. *The Non-Darwinian Revolution: Reinterpreting a Historical Myth*. Baltimore: Johns Hopkins University Press.

———. 2013. *Darwin Deleted: Imagining a World without Darwin*. Chicago: University of Chicago Press.

Boyle, R. [1688] 1966. A disquisition about the final causes of natural things. *The Works of Robert Boyle*. Editor T. Birch, 5: 392–444. Hildesheim: Georg Olms.

Browne, J. 1995. *Charles Darwin: Voyaging*. New York: Knopf.

Brüntrup, G., and L. Jaskolla, editors. 2016. *Panpsychism: Contemporary Perspectives*. Oxford: Oxford University Press.

Budd, S. 1977. *Varieties of Unbelief: Atheists and Agnostics in English Society 1850–1960*. London: Heinemann.

Bullivant, S. B., and M. Ruse, editors. 2013. *The Oxford Handbook of Atheism*. Oxford: Oxford University Press.

———. Forthcoming. *The Cambridge History of Atheism*. Cambridge: Cambridge University Press.

Burtt, E. A. 1932. *The Metaphysical Foundations of Modern Science*. New York: Harcourt, Brace.

Camus, A. [1942] 1955. *The Myth of Sisyphus*. London: Hamish Hamilton.

Chalmers, D. J. 2016. The combination problem for panpsychism. *Panpsychism: Contemporary Perspectives*. Editors G. Brüntrup and L. Jaskolla, 179–214. Oxford: Oxford University Press.

Chambers, R. 1844. *Vestiges of the Natural History of Creation*. London: Churchill.

————. 1846. *Vestiges of the Natural History of Creation*, 5th ed. London: J. Churchill.

Churchland, P. M. 1995. *The Engine of Reason, the Seat of the Soul*. Cambridge, Mass.: MIT Press.

Churchland, P. S. 1986. *Neurophilosophy: Toward a Unified Science of the Mind and Brain*. Cambridge, Mass.: MIT Press.

Clifford, W. K. [1879] 1901. Body and mind. *Lectures and Essays of the Late William Kingdom Clifford*. Editors L. Stephen and F. Pollock, 2: 1–51. London: Macmillan.

Conway Morris, S. 2003. *Life's Solution: Inevitable Humans in a Lonely Universe*. Cambridge: Cambridge University Press.

Cottingham, J. 2003. *On the Meaning of Life*. London: Routledge.

Coyne, J. A. 2015. *Faith versus Fact: Why Science and Religion Are Incompatible*. New York: Viking.

Craig, W. L. 2000. The absurdity of life without God. *The Meaning of Life*. Editor E. D. Klemke, 40–56. New York: Oxford University Press.

Dart, R. 1953. The predatory transition from ape to man. *International Anthropological and Linguistic Review* 1.4: 201–217.

Darwin, C. 1859. *On the Origin of Species by Means of Natural Selection, or the Preservation of Favoured Races in the Struggle for Life*. London: John Murray.

————. 1861. *Origin of Species*, 3rd ed. London: John Murray.

————. 1871. *The Descent of Man, and Selection in Relation to Sex*. London: John Murray.

————. [1887] 1958. *The Autobiography of Charles Darwin, 1809–1882*. Editor Nora Barlow. London: Collins.

Darwin, E. [1794–1796] 1801. *Zoonomia; or, The Laws of Organic Life*, 3rd ed. London: J. Johnson.

————. 1803. *The Temple of Nature*. London: J. Johnson.

Davies, N. B. 1992. *Dunnock Behaviour and Social Evolution*. Oxford: Oxford University Press.

Dawkins, R. 1976. *The Selfish Gene*. Oxford: Oxford University Press.

———. 1983. Universal Darwinism. *Evolution from Molecules to Men*. Editor D. S. Bendall, 403–425. Cambridge: Cambridge University Press.

———. 1986. *The Blind Watchmaker*. New York: Norton.

———. 2003. *A Devil's Chaplain: Reflections on Hope, Lies, Science and Love*. Boston and New York: Houghton Mifflin.

———. 2006. *The God Delusion*. New York: Houghton, Mifflin, Harcourt.

Dawkins, R., and J. R. Krebs. 1979. Arms races between and within species. *Proceedings of the Royal Society of London B* 205: 489–511.

Dennett, D. C. 1992. *Consciousness Explained*. New York: Pantheon.

———. 1995. *Darwin's Dangerous Idea*. New York: Simon & Schuster.

Descartes, R. [1642] 1964. Meditations. *Philosophical Essays*, 59–143. Translator L.I. Lafleur, Indianapolis: Bobbs-Merrill.

———. [1701] 1964. Rules for the direction of the mind. *Philosophical Essays*, 145–236. Translator L. I. Lafleur, Indianapolis: Bobbs-Merrill.

Desmond, A. 1997. *Huxley: From Devil's Disciple to Evolution's High Priest*. New York: Basic Books.

Dickens, C. [1850] 1948. *David Copperfield*. Oxford: Oxford University Press.

Dijksterhuis, E. J. 1961. *The Mechanization of the World Picture*. Oxford: Oxford University Press.

Dixon, M., and G. Radick. 2009. *Darwin in Ilkley*. Stroud, Gloucestershire: History Press.

Dobzhansky, T. 1937. *Genetics and the Origin of Species*. New York: Columbia University Press.

Dostoevsky, F. [1879–80] 2003. *The Brothers Karamazov*. London: Penguin.

Duffy, E. 1992. *The Stripping of the Altars: Traditional Religion in England 1400–1580*. New Haven: Yale University Press.

Edelglass, W., and J. L. Garfield, editors. 2009. *Buddhist Philosophy: Essential Readings*. New York: Oxford University Press.

Fisher, R. A. 1930. *The Genetical Theory of Natural Selection*. Oxford: Oxford University Press.

Ford, E. B. 1964. *Ecological Genetics*. London: Methuen.

Franklin, B. [c. 1790] 2009. *Autobiography and Other Writings*. Oxford: Oxford University Press.

Fry, D. P. 2007. *Beyond War: The Human Potential for Peace*. Oxford: Oxford University Press.

Gould, S. J. 1995. *Dinosaur in a Haystack: Reflections in Natural History*. New York: Harmony.

———. 1999. *Rocks of Ages: Science and Religion in the Fullness of Life*. New York: Ballantine.

Gould, S. J., and E. S. Vrba. 1982. Exaptation—A missing term in the science of form. *Paleobiology* 8: 4–15.

Gray, A. 1876. *Darwiniana*. New York: D. Appleton.

Haldane, J. B. S. 1927. *Possible Worlds and Other Essays*. London: Chatto and Windus.

———. 1932. *The Causes of Evolution*. Ithaca, N.Y.: Cornell University Press.

Hall, A. R. 1954. *The Scientific Revolution 1500–1800: The Formation of the Modern Scientific Attitude*. London: Longman, Green.

———. 1983. *The Revolution in Science, 1500–1750*. London: Longman.

Hamilton, W. D. 1964. The genetical evolution of social behaviour. *Journal of Theoretical Biology* 7: 1–52.

Hardy, T. [1892] 2010. *Tess of the D'Urbervilles*. London: Collins.

———. 1994. *Collected Poems*. Ware: Wordsworth Poetry Library.

Harrington, A. 1996. *Reenchanted Science: Holism in German Culture from Wilhelm II to Hitler*. Princeton: Princeton University Press.

Harrison, P. 2009. *The Fall of Man and the Foundations of Science*. Cambridge: Cambridge University Press.

Harvey, P. 1990. *An Introduction to Buddhism: Teachings, History and Practices*. Cambridge: Cambridge University Press.

Heidegger, M. 1959. *An Introduction to Metaphysics*. New Haven: Yale University Press.

Herschel, J. F. W. 1830. *Preliminary Discourse on the Study of Natural Philosophy*. London: Longman, Rees, Orme, Brown, Green, and Longman.

Hick, J. 1973. *God and the Universe of Faiths: Essays in the Philosophy of Religion*. New York: St. Martin's Press.

———. 1980. *God Has Many Names*. Philadelphia: Westminster Press.

———. 2005. *An Autobiography*. London: Oneworld Publications.

Hobbes, T. [1651] 1982. *Leviathan*. Harmondsworth: Penguin.

Hodge, C. 1874. *What Is Darwinism?* New York: Scribner's.

Hrdy, S. B. 1981. *The Woman That Never Evolved*. Cambridge, Mass.: Harvard University Press.

———. 1999. *Mother Nature: A History of Mothers, Infants, and Natural Selection*. New York: Pantheon Books.

Hull, D. L. 1986. On human nature. *PSA: Proceedings of the Biennial Meeting of the Philosophy of Science Association* 2: 3–13.

Hume, D. [1739–40] 1978. *A Treatise of Human Nature*. Oxford: Oxford University Press.

———. [1777] 1902. *Enquiries Concerning the Human Understanding, and Concerning the Principles of Morals*. Oxford: Oxford University Press.

———. [1779] 1963. Dialogues concerning natural religion. *Hume on Religion*. Editor R. Wollheim, 93–204. London: Fontana.

Humphreys, P. 2016. *Emergence*. New York: Oxford University Press.

Huxley, J. S. 1927. *Religion without Revelation*. London: Ernest Benn.

———. 1942. *Evolution: The Modern Synthesis*. London: Allen and Unwin.

———. 1943. *TVA: Adventure in Planning*. London: Scientific Book Club.

———. 1948. *UNESCO: Its Purpose and Its Philosophy*. Washington, D.C.: Public Affairs Press.

Huxley, L. 1900. *The Life and Letters of Thomas Henry Huxley*. London: Macmillan.

Huxley, T. H. [1860] 1893. The origin of species. *Darwiniana*. Editor T. H. Huxley, 22–79. London: Macmillan.

———. [1893] 2009. *Evolution and Ethics*, edited with introduction by Michael Ruse. Princeton: Princeton University Press.

Irenaeus of Lyons. [c. 180] 2018. *Against Heresies*. Translators A. Roberts and W. Rambaut. Ennis, Eire: Dalcassian.

Ismael, J., and J. Schaffer. 2016. Quantum holism: Nonseparability as common ground. *Synthese* (September): 1–30.

James, W. 1902. *The Varieties of Religious Experience: A Study in Human Nature*. New York: Longman.

Jerison, H. 1973. *Evolution of the Brain and Intelligence*. New York: Academic Press.

John Paul II. 1997. The pope's message on evolution. *Quarterly Review of Biology* 72: 377–383.

———. 1998. *Fides et Ratio: Encyclical Letter of John Paul II to the Catholic Bishops of the World*. Vatican City: L'Osservatore Romano.

Kant, I. [1790] 2000. *Critique of the Power of Judgment*. Editor P. Guyer. Cambridge: Cambridge University Press.

Kauffman, S. A. 1993. *The Origins of Order: Self-Organization and Selection in Evolution*. Oxford: Oxford University Press.

Kepler, J. [1619] 1997. *The Harmony of the World*. Translators E. J. Aiton, A. M. Duncan, and J. V. Field. Philadelphia: American Philosophical Society.

Kierkegaard, S. 1944. *Concluding Unscientific Postscript*. Translator D. F. Swenson. Princeton: Princeton University Press.

Kimler, W., and M. Ruse. 2013. Mimicry and camouflage. *The Cambridge Encyclopedia of Darwin and Evolutionary Thought*. Editor M. Ruse, 139–145. Cambridge: Cambridge University Press.

Kitcher, P. 2007. *Living with Darwin: Evolution, Design, and the Future of Faith*. New York: Oxford University Press.

———. 2014. *Life after Faith: The Case for Secular Humanism*. New Haven: Yale University Press.

Krauss, L. M. 2012. *A Universe from Nothing*. New York: Atria.

Kuhn, T. 1957. *The Copernican Revolution*. Cambridge, Mass.: Harvard University Press.

———. 1962. *The Structure of Scientific Revolutions*. Chicago: University of Chicago Press.

———. 1993. Metaphor in science. In *Metaphor and Thought*, 2nd ed. Editor Andrew Ortony, 533–542. Cambridge: Cambridge University Press.

Leibniz, G. W. F. [1714] 1989. *Philosophical Essays*. Editors R. Ariew and D. Garber. Indianapolis: Hackett.

Lewis, C. S. 1955. *Surprised by Joy: The Shape of My Early Life*. London: Geoffrey Bles.

Lewontin, R. C. 1991. *Biology as Ideology: The Doctrine of DNA*. Toronto: Anansi.

———. 2000. *The Triple Helix: Gene, Organism, and Environment*. Cambridge, Mass.: Harvard University Press.

Lorenz, K. 1966. *On Aggression*. London: Methuen.

Lucretius. 1950. *Of the Nature of Things*. Translator W. E. Leonard. London: Dutton.

MacArthur, R. H., and E. O. Wilson. 1967. *The Theory of Island Biogeography*. Princeton: Princeton University Press.

MacCulloch, D. 2004. *The Reformation: A History*. New York: Viking.

Malthus, T. R. [1826] 1914. *An Essay on the Principle of Population*, 6th ed. London: Everyman.

Maynard Smith, J. 1964. Group selection and kin selection. *Nature* 201: 1145–1147.

Mayr, E. 1942. *Systematics and the Origin of Species*. New York: Columbia University Press.

McGinn, C. 2000. *The Mysterious Flame: Conscious Minds in a Material World*. New York: Basic Books.

McShea, D. W., and R. Brandon. 2010. *Biology's First Law: The Tendency for Diversity and Complexity to Increase in Evolutionary Systems*. Chicago: University of Chicago Press.

Merchant, C. 1980. *The Death of Nature: Women, Ecology, and the Scientific Revolution: A Feminist Reappraisal of the Scientific Revolution*. Scranton, Pa.: HarperCollins.

Meredith, G. 1870. In the woods. *Fortnightly Review* 8: 179–183.

Metz, T. 2013. *Meaning in Life*. New York: Oxford University Press.

Mitchell, D. W., and S. H. Jacoby. 2014. *Buddhism: Introducing the Buddhist Experience, Third Edition*. New York: Oxford University Press.

Moore, A. 1890. The Christian doctrine of God. *Lux Mundi*. Editor C. Gore, 57–109. London: John Murray.

Moore, G. E. 1903. *Principia Ethica*. Cambridge: Cambridge University Press.

———. 1925. A defence of common sense. *Contemporary British Philosophy*. Editor J. H. Muirhead, Vol. 2, 193–223. London: Allen and Unwin.

Nagel, T. 2012. *Mind and Cosmos: Why the Materialist Neo-Darwinian Conception of Nature Is Almost Certainly False*. New York: Oxford University Press.

Newman, J. H. 1973. *The Letters and Diaries of John Henry Newman, XXV*. Editors C. S. Dessain and T. Gornall. Oxford: Clarendon Press.

Nietzsche, F. [1895] 1990. *The Anti-Christ*. London: Penguin.

Paley, W. [1794] 1819. *A View of the Evidences of Christianity (Collected Works III)*. London: Rivington.

———. [1802] 1819. *Natural Theology: or, Evidences of the Existence and Attributes of the Deity (Collected Works IV)*. London: Rivington.

Peterson, M., and M. Ruse. 2016. *Science, Evolution, and Religion: A Debate about Atheism and Theism*. Oxford: Oxford University Press.

Pinker, S. 2011. *The Better Angels of Our Nature: Why Violence Has Declined*. New York: Viking.

Plantinga, A. 2000a. Pluralism: A defense of religious exclusivism. *The Philosophical Challenge of Religious Diversity*. Editors K. Meeker and P. Quinn, 172–192. New York: Oxford University Press.

———. 2000b. *Warranted Christian Belief*. Oxford: Oxford University Press.

Pleins, J. D. 2003. *When the Great Abyss Opened: Classic and Contemporary Readings of Noah's Flood*. Oxford: Oxford University Press.

Provine, W. B. 1971. *The Origins of Theoretical Population Genetics*. Chicago: University of Chicago Press.

Quinn, P. L. 1978. *Divine Commands and Moral Requirements*. Oxford: Clarendon Press.

Reiss, M., and M. Ruse. Forthcoming. *The New Biology: History, Philosophy and Implications*. Cambridge, Mass.: Harvard University Press.

Richards, R. J. 1987. *Darwin and the Emergence of Evolutionary Theories of Mind and Behavior*. Chicago: University of Chicago Press.

———. 2003. *The Romantic Conception of Life: Science and Philosophy in the Age of Goethe*. Chicago: University of Chicago Press.

Richards, R. J., and M. Ruse. 2016. *Debating Darwin*. Chicago: University of Chicago Press.

Ruse, M. 1973. *The Philosophy of Biology*. London: Hutchinson.

———. 1979a. *The Darwinian Revolution: Science Red in Tooth and Claw*. Chicago: University of Chicago Press.

———. 1979b. *Sociobiology: Sense or Nonsense?* Dordrecht: Reidel.

———. 1986. *Taking Darwin Seriously: A Naturalistic Approach to Philosophy*. Oxford: Blackwell.

———, editor. 1988a. *But Is It Science? The Philosophical Question in the Creation/Evolution Controversy*. Buffalo: Prometheus.

———. 1988b. *Homosexuality: A Philosophical Inquiry*. Oxford: Blackwell.

———. 1996. *Monad to Man: The Concept of Progress in Evolutionary Biology*. Cambridge, Mass.: Harvard University Press.

———. 2001a. *Can a Darwinian Be a Christian? The Relationship between Science and Religion*. Cambridge: Cambridge University Press.

———. 2001b. Evolutionary naturalism. *The Nature and Limits of Human Understanding (Gifford Lectures, 2001)*. Editor A. Sanford, 109–162. Edinburgh: T. and T. Clark.

———. 2003. *Darwin and Design: Does Nature Have a Purpose?* Cambridge, Mass.: Harvard University Press.

———. 2005. *The Evolution-Creation Struggle*. Cambridge, Mass.: Harvard University Press.

———. 2006. *Darwinism and Its Discontents*. Cambridge: Cambridge University Press.

———. 2008. *Charles Darwin*. Oxford: Blackwell.

———. 2010. *Science and Spirituality: Making Room for Faith in the Age of Science*. Cambridge: Cambridge University Press.

———. 2012. *The Philosophy of Human Evolution*. Cambridge: Cambridge University Press.

———. 2013. *The Gaia Hypothesis: Science on a Pagan Planet*. Chicago: University of Chicago Press.

———. 2015. *Atheism: What Everyone Needs to Know*. Oxford: Oxford University Press.

———. 2016. *Evolution and Religion: A Dialogue*, 2nd ed. Lanham, Md.: Rowman and Littlefield.

———. 2017a. *Darwinism as Religion: What Literature Tells Us about Evolution*. Oxford: Oxford University Press.

———. 2017b. *On Purpose*. Princeton: Princeton University Press.

———. 2017c. Social Darwinism. *On Human Nature: Biology, Psychology, Ethics, Politics, and Religion*. Editors M. Tibayrence and F. J. Ayala, 651–660. Amsterdam: Academic Press.

———. 2018a. *The Darwinian Revolution: What Philosophers Should Know*. Cambridge: Cambridge University Press.

———. 2018b. *The Problem of War: Darwinism, Christianity, and Their Battle to Understand Human Conflict*. Oxford: Oxford University Press.

Ruse, M., and R. J. Richards, editors. 2017. *The Cambridge Handbook of Evolutionary Ethics*. Cambridge: Cambridge University Press.

Ruse, M., and E. O. Wilson. 1986. Moral philosophy as applied science. *Philosophy* 61: 173–192.

Sartre, J.-P. 1948. *Existentialism and Humanism*. Brooklyn, N.Y.: Haskell House Publishers.

Schopenhauer, A. [1851] 2008. Parega and paralipomena. *The Meaning of Life: A Reader*. Editors E. D. Klemke and S. M. Cahn, 45–54. New York: Oxford University Press.

Scruton, R. 2017. *On Human Nature*. Princeton: Princeton University Press.

Sebright, J. 1809. *The Art of Improving the Breeds of Domestic Animals in a Letter Addressed to the Right Hon. Sir Joseph Banks, K.B.* London: privately published.

Simpson, G. G. 1944. *Tempo and Mode in Evolution*. New York: Columbia University Press.

Skilton, A. 2013. Buddhism. *The Oxford Handbook of Atheism*. Editors S. Bullivant and M. Ruse, 337–350. Oxford: Oxford University Press.

Spencer, H. 1857. Progress: Its law and cause. *Westminster Review* 67: 244–267.

———. 1879. *The Data of Ethics*. London: Williams and Norgate.

Stebbins, G. L. 1950. *Variation and Evolution in Plants*. New York: Columbia University Press.

Steiner, R. [1914] 2005. *Occult Science: An Outline*. Forest Row, Sussex: Rudolf Steiner Press.

Stenger, V. J. 2011. *The Fallacy of Fine-Tuning: Why the Universe Is Not Designed for Us*. Buffalo: Prometheus.

Swinburne, R. 2004. *The Existence of God*, 2nd ed. Oxford: Clarendon Press.

Taylor, C. 1975. *Hegel*. Cambridge: Cambridge University Press.

———. 2007. *A Secular Age*. Cambridge, Mass.: Harvard University Press.

Tolstoy, L. [1882] 2008. My confession. *The Meaning of Life: A Reader*. Editors E. D. Klemke and S. M. Cahn, 7–16. New York: Oxford University Press.

Trivers, R. L. 1971. The evolution of reciprocal altruism. *Quarterly Review of Biology* 46: 35–57.

Tuttle, R. H. 2014. *Apes and Human Evolution*. Cambridge, Mass.: Harvard University Press.

Ward, Mrs. H. 1888. *Robert Elsmere*. London: Macmillan.

Ward, K. 2004. *What the Bible Really Teaches: A Challenge for Fundamentalists*. London: SPCK.

Weinberg, S. 1977. *The First Three Minutes: A Modern View of the Origin of the Universe*. New York: Basic Books.

———. 1999. A designer universe? *New York Review of Books* 46, 16: 46–48.

Westfall, R. S. 1980. *Never at Rest: A Biography of Isaac Newton*. Cambridge: Cambridge University Press.

Whewell, W. 1840. *The Philosophy of the Inductive Sciences*. London: Parker.

Williams, B. 1973. The Makropulos case: Reflections on the tedium of immortality. *Problems of the Self*. Cambridge: Cambridge University Press.

Wilson, E. O. 1975. *Sociobiology: The New Synthesis*. Cambridge, Mass.: Harvard University Press.

———. 1978. *On Human Nature*. Cambridge, Mass.: Harvard University Press.

———. 1984. *Biophilia*. Cambridge, Mass.: Harvard University Press.

———. 1992. *The Diversity of Life*. Cambridge, Mass.: Harvard University Press.

———. 2002. *The Future of Life*. New York: Vintage Books.

———. 2006. *The Creation: A Meeting of Science and Religion*. New York: Norton.

———. 2014. *The Meaning of Human Existence*. New York: Liveright.

Wittgenstein, L. 1965. A lecture on ethics. *Philosophical Review* 74: 3–12.

Wright, S. 1931. Evolution in Mendelian populations. *Genetics* 16: 97–159.

———. 1932. The roles of mutation, inbreeding, crossbreeding and selection in evolution. *Proceedings of the Sixth International Congress of Genetics* 1: 356–366.

INDEX

Absolute Spirit (Hegel), 130
"accommodationism" (Gould), 59
Adam (Biblical), 57, 89
adaptations, 33, 115, 119–20
Against Heresies (Irenaeus of Lyons), 58
agency (direct Divine), 50
agnosticism, 32, 51, 169–71
altruism, 148–49, 151
anatomy, 40
Anscombe, Elizabeth, 64
Anselm, Saint, 67, 84
anthropic principle, 71
ants, 76–77
Apollonia, Saint, 13–14
Aquinas, Saint Thomas, 17, 66, 69, 82, 84
Ardley, Robert, 156
Aristotle, 22–24, 39–40, 82
"arms race," 115, 117, 120–21
aseity, 67
atheism, 32, 133
    Darwinism and, 49–50
    "intellectually fulfilled" (Dawkins), 52
atonement (substitutionary), 50, 57
Augustine, Saint, 19, 20, 56, 57–59, 82–83, 156
Autobiography (Darwin), 41

Bacon, Francis, 22–24
bank, entangled (Darwin), 40–41
barnacles, 45
Barth, Karl, 69
Bates, Henry Walter, 45
bats, 27
Becket, Saint Thomas à, 13, 14
Benatar, David, 6
Bergson, Henri, 126
Berkeley, George, 97
Bible, 82, 137–38
    Augustine and, 156
    authenticity of, 30
    Church authority and, 18–20
    Erasmus and, 16
    Higher Criticism and, 50
    science and, 55–57
    vernacular, 19
Big Bang theory, 63
biogeography, 40
biology, 19–21, 134–35, 136–53.
        See also Christianity; culture;
        Darwinism; evolution
    final causes and, 30–31
    "First Law" of (McShea and Brandon), 119
    human nature and, 142
    molecular, 48–49
    physical sciences and, 31–32

"biophilia" (Wilson), 125
Blake, William, 109–10
Bonheoffer, Dietrich, 78, 84
Boyle, Robert, 26, 27–28
Brandon, Robert, 119–20
Brewster, David, 105
Bruno, Giordano, 21
Buckland, William, 56
Buddha, 86–90
Buddhism, 30, 81, 85–96, 97
Burns, Robert, 61

Calvin, John, 18–20
Calvinism, 96
Camus, Albert, 6, 9, 53
Catholicism, 12–13, 82, 145
causation (efficient and final), 22–24,
        30–31, 32–33
Chambers, Robert, 104–5
characteristics, inherited
        (Lamarck), 33
Christ, Jesus, 4, 14, 21, 80, 86
    Augustine and, 57–58
    God and, 12–13
    as perfect love, 58–60
    as Redeemer, 108
    Ward and, 84–85
Christianity, 76–86, 89, 91–93, 97,
        113. See also atheism; Buddhism;
        consciousness; deism; evolution;
        Judaism; theism
    Darwin and, 51
    Eastern, 58
    God and, 12–13
    Hall and, 24
    heliocentrism and, 25
    Hick and, 60

    Huxley and, 106–7
    Kierkegaard and, 18
    Reformation and, 18–19
    Romanticism and, 130
    roots of, 82
    saving, 79
    science and, 59–60, 66–75, 105–6
    versions of, 19, 104
    Wilson and, 108–9
chromosomes, 47
Churchland, Pat and Paul, 65
Clifford, William Kingdom, 165
Coleridge, Samuel, 112
competition, 115
Complementarity, 64
complexity, 119–20
consciousness, 64, 90, 130,
        151–53, 163–69
    Christianity and, 66, 67
    value and, 167–69
Constable, John, 80, 82
consubstantiation, 20
convergence, 118–19
cooperation, 42–43
Copernican Revolution, 21, 39
Corpus Christi, 15
Cosi Fan Tutte (Mozart/ da Ponte),
        154, 167
cosmos, 24, 26
Coyne, Jerry, 48, 54–55
Craig, William Lane, 4
creativity, 153
Critique of the Power of Judgment
        (Kant), 32
culture, 80, 121, 137–43, 144, 149.
        See also biology; "exaption"
        (Gould)

da Ponte, Lorenzo, 154

Dart, Raymond, 156

Darwin, Charles, 7–10, 32–45, 47–48, 109, 113–17

Darwin, Erasmus, 32–34, 104

Darwinism. *See also* Existentialism, Darwinian
Aristotle and, 39–40
defined, 98–100
as religion, 97–132

David Copperfield (Dickens), 72–75, 152

Dawkins, Richard, 67, 92, 99–100. *See also* "selfish genes" (Dawkins)

De rerum natura (Lucretius), 17–18

death, 3, 6, 15

Decline and Fall (Waugh), 5

deism, 32–33, 34, 51

Demiurge, 23

Democritus, 17–18

Dench, Dame Judi, 15

Dennett, Daniel, 64, 125

Descartes, René, 22–24, 26, 167

Descent of Man, The, and Selection in Relation to Sex (Darwin), 7, 42–43, 117, 148

design, argument from, 26–27, 28–29, 39–40, 51, 55, 89–90

determinism (biological), 135–36

Dialogues Concerning Natural Religion (Hume), 28–29

Dickens, Charles, 159

Diderot, Denis, 28

differentiation, 116

diversity, 119–20, 125

Dobzhansky, Theodosius, 46–48

Dostoevsky, Fyodor, 78

drift, genetic, 48

dualism (Cartesian), 164

dukkha, 89

duty, 146
to be good, 20–21, 67
evolution and, 147–50
to God, 109–10
Hume and, 146

"dynamic equilibrium" (Spencer), 112

Easter, 14

Eightfold Path, 90

embryology, 40

emergentism, 164

entanglement, quantum, 165, 166

"epigenetics," 48–49

EQ (encephalization quotient), 117

Erasmus of Rotterdam, 16, 19

eternity, 161–64, 169

ethics, 64, 108, 123, 125

eugenics, 123–24

Euthyphro problem, 67

Eve, 89

evil, 76–78

"evo- devo," 48

evolution, 36. *See also* progress
Christianity and, 101–2
deism and, 32–34
God as designer and, 55
organic progressive, 104–11
post- Darwinian, 44–49
spiritualism and, 42

Evolution: The Fossils Say No! (Gish), 168

"exaptation" (Gould), 153

existence, struggle for, 5–6, 35–38, 51–52, 53

Existentialism, 133–36

facts, 126, 128, 129
faith (*sola fidelis*), 18, 83
    belief and, 69–70
    fact and, 55–56
    reason and, 69–70
*Faith versus Fact: Why Science and
    Religion Are Incompatible*
    (Coyne), 54
Fall of Man, 20
family. *See* humans, sociality and
Farn, Albert Bridges, 46
Fisher, Ronald, 46–48
Flood, Noah's, 56
Ford, E. B., 46–47
Form of the Good (Plato), 17, 23,
    84, 159
Frank, Anne, 78, 84
Frankenstein (movie), 135
Franklin, Benjamin, 32, 36
Freudianism, 136
friendship, Aristotelian. *See* humans,
    sociality and
Future of Life, The (Wilson), 125

Galton, Francis, 41
Gautama, Siddhārtha, 86
gene, classic theory of the (Morgan), 47
genetics, 44, 147
Genetics and the Origin of Species
    (Dobzhansky), 48
Gish, Duane T., 168
God. *See also* design, argument from
    Aristotle and, 24–25
    belief, decline of, 9
    causal chain and, 66
    as Creator, 3, 12, 59–61, 91–92, 104
        defining, 80–81
        direct insight into, 79
        existence of, 133–34
        as First Mover, 25–26
        intention and, 22
        Judaism and, 83
        Old Testament, 4, 100
        as a retired engineer, 27
        teleology and, 26
        value and, 4–5
God Delusion, The (Dawkins), 91
Good, the (Plato), 17
Good Friday, 14
Goodall, Jane, 58
Gould, Stephen Jay, 59–61, 135–36, 153
Grace, 19
Gray, Asa, 115, 122

Haldane, J. B. S., 47, 153
Hall, Rupert, 24, 26
Hamilton, William, 147
Hardy, Thomas, 51–53, 88, 97,
    130–31, 153–54
Harrison, Peter, 20
Hay Wain, The (Constable), 80
Hegel, G. W. F., 130
Heidegger, Martin, 63
heliocentrism, 5–6, 21, 25, 39
heredity, 43, 47
Herschel, John F. W., 37
Hick, John
    Christian conversion and, 60, 75,
        84, 169–70
    transcendent being and, 92
Higher Criticism, the, 30, 50
Himmler, Heinrich, 76, 78, 122

Hinduism, 30
Hitler, Adolf, 68, 78
Hobbes, Thomas, 12
Hodge, Charles, 49
holism, 151, 166
homologies, 45
Hrdy, Sarah, 139
"human condition," 6
humanism, 107, 108, 128
humans, 41–43, 116–19,
        121–23, 125–26
    human nature, 141–42, 158
    as killer apes, 154–58
    machine metaphor of, 26
    sociality and, 143, 147–49
Hume, David, 146–49
    Hume's Law, 126–27, 129, 168–69
    skepticism and, 1
Huxley, Julian, 102–3, 106–7, 111,
        112, 124, 126
Huxley, Thomas Henry, 100–2,
        109–10, 111
Hymenoptera, 147

incarnation, 93–96
inductions, consilience of
        (Whewell), 37
insects, social, 61–62
intelligence, human, 139
intention, 40
Irenaeus of Lyons, 58
Islam, 80
is/ought barrier, 128

James, Saint, 14
James, William, 3–4, 74

John Paul II, Pope, 69–71, 96, 145, 163
Judaism, 80, 83

Kant, Immanuel, 30–31, 32–33, 148
Kauffman, Stuart, 119
Kepler, Johannes, 25–26
Kierkegaard, Søren, 51, 69
Kitcher, Philip, 128–29
Kuhn, Thomas, 61

lactose intolerance, 137–39, 167
"Lay Sermons" (T. H. Huxley), 102
Leibniz, Gottfried, 64
Lewontin, Richard, 149
Lichtenstein, Roy, 154
life
    meaning of, and natural
        selection, 6–7
    tree of, 40
Life after Faith: The Case for Secular
        Humanism (Kitcher), 128
Lorenz, Konrad, 155–56
Luther, Martin, 15, 18–20

Magisterium (Gould), 59
Malthus, Reverend Thomas
        Robert, 35–36
Māra, 86
Mary, Mother of God, 13, 20
materialism, 61
matter in motion, 23
Matthew, Saint, 161
Maundy Thursday, 14
Mayr, Ernst, 47
McGinn, Colin, 65, 168
McGruff, Scruffy, 162, 167–68

McShea, Daniel, 119
meaning, 62–66
    Aristotle and, 156–57
    Buddhism and, 92–93
    Christianity and, 66, 68, 92–93
    evolutionists and, 110
    God- centered. *See* Christianity
    loss of, 49–53
    objectivism and, 98–111
    Providence and, 109
    soul- centered, 85–96
    Ultimate, 65, 169
Meno (Plato), 93
metaethics, 64
metaphors
    Darwin and, 61–62
    paradigms and, 61
    root, 22–27
metaphysics, fundamental question of
    (Heidegger), 63
Metz, Thaddeus, 85
Middle Ages, 11–16
mimicry (Bates and Müller), 44–46
miracles, 28–29, 51, 60
Modern Synthesis, The (J. Huxley), 103
Moore, Aubrey, 49–50
Moore, G. E., 79, 126, 149
morality, 42, 63–64, 127, 147–51
    Christianity and, 66, 67
    Darwin and, 149–50
Morgan, Thomas Hunt, 47
"morphological space" (Morris), 118
morphology, 40
Morris, Simon Conway, 118–19
Mozart, Wolfgang Amadeus, 154
Müller, Fritz, 45

mutations (random variation), 47, 115,
    125. *See also* variation

Nagel, Thomas, 122
National Socialism, 151
nativism, 136
natural law theory, 145
*Natural Theology: or, Evidences of the
    Existence and Attributes of the
    Deity* (Paley), 30
naturalism, 103, 126–32
    methodological (Boyle), 28
    as non- theism, 9
"naturalistic fallacy" (Moore), 126
Neo- Darwinism, 48
New Atheism, 23
Newman, John Henry, 51, 70
Newton, Isaac, 27
Nicomachean Ethics (Aristotle), 143
Nietzsche, Friedrich, 58
nirvana (nibbana), 88, 89–93
no- self (anatta, anatman), 88
nothingness, 6
nurturism, 136

*On the Origin of Species by Means
    of Natural Selection, or the
    Preservation of Favoured Races in
    the Struggle for Life* (Darwin)
    Bates and, 45
    Coyne and, 54–55
    Darwinism and, 98–99
    Darwin's deism and, 51
    Dobzhansky and, 48
    Homo sapiens and, 41, 116
    Huxley and, 135

importance of, 7
progress, selection and, 115–16
Sedgwick and, 56
Spencer and, 106
Wallace and, 37
organisms, 23–24, 27–31
organs, vestigial, 40
Orwell, George, xi
Oxford Movement, 51

pacifism, 90
paganism, 81–82
paleontology, 40
Paley, Archdeacon William, 30,
    39–40, 75
Palm Sunday, 14
panpsychism, 65, 164–69
parenting, human, 140–42
Parry, Hubert, 110
Pascal, Blaise, 5
Paul, Saint, 13, 14, 75, 91, 162, 163
peacock, 43
Peaseblossom, 58, 81
Peter, Saint, 13, 14, 91
Phaedo (Plato), 90
pigeons, 37
Pinker, Steven, 157–58
Plantinga, Alvin, 55–56, 70,
    78–80, 82, 96
Plato, 25, 27, 39–40, 82, 90, 149.
    See also Form of the Good (Plato)
Plotinus, 17, 82
pluralism, 21, 79–82, 92, 97–98
Popper, Karl, 20
printing, 19
Prodigal Son, 83–84

progress, 104–26. See also evolution
Protestantism, 82
Providence, 104–10
Purgatory, 15–16, 19, 87
purpose (intention), and
    causation, 22

reason, 70–71
    Christianity and, 75
    faith and, 69–70
reductionism, biological, 150
Reformation, 18–21
reincarnation, 87
religion
    Darwin and, 51, 56–57, 134
    human fate and, 3
    meaning and, 12–15
    medieval, 8, 11–16.
        See also Christianity
    objectivism and, 98–111
    science and, 54–66
Religion without Revelation
    (J. Huxley), 102
"religious exclusivism," 79, 80
Renaissance, 16–17, 23–24, 29
reproduction, 145
Republic (Plato), 158–61
res cogitans, 23, 88–89, 164
res extensa, 23, 164
Resurrection, 50, 60, 162
Revelation, 71, 163
Robert Elsmere (Ward), 102
Romanticism (German), 112, 122,
    126, 130
Rudge, Martin, 55
Russell, Bertrand, 2

saltations, 77
salvation (personal), 91, 108
Sartre, Jean- Paul, 133–35
Schelling, Friedrich, 112
Scholl, Sophie, 78, 84, 149–50
Schopenhauer, Arthur, 5–6
science, 54–66, 163
Scientific Revolution, 20–22, 23
Scruton, Roger, 150, 152–53
Sedgwick, Adam, 56, 105
selection
    "kin" (J. M. Smith), 147
    k-selection, 77
    natural, 34–37, 38–39, 40,
        44–45, 46–49, 54–55, 77,
        115, 125–26, 130–31
    non- Darwinian reading of, 150
    r- selection, 77
    sexual, 43, 153
self (Buddhism), 88
"selfish genes" (Dawkins), 43, 150–51
self- organization (Kauffman), 120
Sepkoski, Jack, 115, 168
sexuality, human, 139
Shakespeare, William, 1, 4,
    131–32, 154
Simpson, George Gaylord, 47
sin, original, 20–21, 57–58, 95–96
Smith, John Maynard, 147
sociality, human, 142–47,
    154–55, 156–57
Socrates, 7
specialization, 116, 137
Spencer, Herbert, 106–7, 111–13,
    123–24, 125–26
Spinoza, Baruch, 30

Stebbins, G. Ledyard, 47
Steiner, Rudolf, 93
Stephen, Saint, 13
suicide, 6
Swinburne, Richard, 83
systematics, 40, 45

Taung Baby (*Australopithecus
    afarensis*), 156
Taylor, Charles, 61, 63–64
"tedium of immortality" (Williams), 161
teleology, 26, 40
Tess of the D'Urbervilles (Hardy),
    88, 153–54
theism, 15, 83
    deism and, 32–33
    non- theism (naturalism) and, 9
    postulates of, 83
theology (natural and revealed), 23,
    28–29, 30, 51
Thirty- Nine Articles, 71
Thomas, Saint, 69
Thrasymachus, 159
Timaeus (Plato), 23, 40
Tolstoy, Leo, 2
Tractatus (Wittgenstein), 7–9
Trollope. Anthony, 15
truth, 18–21. *See also* Bible
TVA (Tennessee Valley
    Authority), 124

Uncertainty Principle
    (Heisenberg), 164
UNESCO (United nations
    Educational, Scientific and
    Cultural Organization), 107

Unitarianism, 81
Unmoved Mover, the (Aristotle), 17, 23, 27, 52, 122

value, 4–5, 123, 124–26, 128, 130
    causation and, 22
    consciousness and, 167–69
    meaning and, 9
van Gogh, Vincent, 80, 82
variation, 115, 137. *See also* mutations (random variation)
vegetarianism, 90
Vermeer, Johannes, 59
Vestiges of the Natural History of Creation (Chambers), 104–5
View of the Evidences of Christianity, A (Paley), 30
violence, 155–58

Wallace, Alfred Russel, 37, 42, 43
Ward, Keith, 84–85, 95–96
Ward, Mrs. Humphrey (Mary Augusta Ward), 102–3
Watts, Reverend Isaac, 104

Waugh, Evelyn, 5
Wedgwood, Emma, 113
Weinberg, Steven, 54, 61, 66, 71
Wheeler, W. H., 113
Whewell, William, 37, 62, 105
"Why Evolution Is True" (Coyne), 49
will, free, 76
Williams, Bernard, 161
Wilson, Edward O.
    biological progress and, 107–9
    "biophilia" and, 124–26
    "facts" and, 128
    mythological evolution and, 103
    Ruse and, 162
    Spencer and, 112–13, 123
    Tyler Prize and, 111
Wittgenstein, Ludwig, 8, 63
Wright, Sewall, 46–48

York Mystery Plays, 15

ZFEL (zero- force evolutionary law), 119–20, 122